Data-driven Modeling and Optimization of Multi-Energy Systems

Datengetriebene Modellierung und Optimierung von Multi-Energiesystemen

Von der Fakultät für Maschinenwesen der
Rheinisch-Westfälischen Technischen Hochschule Aachen
zur Erlangung des akademischen Grades eines Doktors
der Ingenieurwissenschaften genehmigte Dissertation

vorgelegt von

Andreas Kämper geb. Hüttermann

Berichter: Univ.-Prof. Dr.-Ing. André Bardow
Univ.-Prof. Dr.-Ing. Dirk Müller

Tag der mündlichen Prufung: 7. März 2023

Diese Dissertation ist auf den Internetseiten der Universitätsbibliothek online verfügbar.

Aachener Beiträge zur Technischen Thermodynamik Band 41

Andreas Kämper
Data-driven Modeling and Optimization of Multi-Energy Systems

Datengeschriebene Modellierung und Optimierung von Multi-Energiesystemen

ISBN: 978-3-95886-488-7

Bibliografische Information der Deutschen Bibliothek
Die Deutsche Bibliothek verzeichnet diese Publikation in der Deutschen Nationalbibliografie; detaillierte bibliografische Daten sind im Internet über http://dnb.ddb.de abrufbar.

Herstellung & Vertrieb:

1. Auflage 2023
© Wissenschaftsverlag Mainz GmbH - Aachen
Süsterfeldstr. 83, 52072 Aachen
Tel. 0241 / 87 34 34 00
www.Verlag-Mainz.de

ISSN: 2198-4832

Satz: nach Druckvorlage des Autors
Umschlaggestaltung: Druckerei Mainz

printed in Germany
D82 (Diss. RWTH Aachen University, 2023)

Danksagung

Die vorliegende Arbeit stellt das Ergebnis meiner Tätigkeit als wissenschaftlicher Mitarbeiter am Lehrstuhl für Technische Thermodynamik der RWTH Aachen sowie am Energy & Process Systems Lab der ETH Zürich dar. Diese Arbeit wäre ohne die Hilfe und Unterstützung vieler Menschen nicht möglich gewesen.

Zunächst möchte ich mich bei meinem Doktorvater Prof. Dr.-Ing. André Bardow bedanken für die inspirierende Betreuung, das entgegengebrachte Vertrauen und die gewährten Möglichkeiten der Weiterentwicklung, sowohl fachlich als auch persönlich. Ebenfalls bedanken möchte ich mich bei Prof. Dr.-Ing Dirk Müller für die Übernahme des Korreferats und bei Prof. Dr.-Ing. Dirk Abel für Übernahme des Prüfungsvorsitzes.

Ein weiterer Dank gilt meinen Kolleg:innen und Freunden am LTT und am EPSE, die mich mit hilfreichen Ratschlägen und kritischen Fragen unterstützt haben. Ihre wertvollen Beiträge haben dazu beigetragen, meine Arbeit auf ein höheres Niveau zu heben. Dazu zählen auch zahlreiche Student:innen, die diese Arbeit mit ihrem großen Engagement unterstützt haben, allen voran Roman, Hendrik, Alexander und Philipp. Der Dank an die Kolleg:innen und Student:innen geht jedoch über den Beitrag zu dieser Arbeit hinaus. Der einzigartige Zusammenhalt und die anregenden Diskussionen haben mich persönlich geprägt und mir wertvolle Erfahrungen vermittelt, die mich mein weiteres Leben begleiten werden.

Im besonderen möchte ich Dr. Ludger Leenders danken, der mich über die gesamte Zeit meiner Promotion begleitet hat. Ich danke Dir für die unbegrenzte fachliche Unterstützung und die vielen persönlichen Gespräche. Ebenfalls ein besonderer Dank gilt Dr. Johannes Schilling für die bedingungslose Hilfe in vielen Momenten ohne einen Gedanken an eigene Interessen. Ich danke Euch für die uns verbindende Freundschaft.

Schließlich möchte ich meiner Familie und weiteren Freunden danken, die mich durch ihren moralischen Beistand und ihre Geduld in dieser Zeit begleitet haben. Ohne ihre Liebe und Unterstützung hätte ich diese Herausforderung nicht bewältigen können. Meinen Eltern danke ich besonders dafür, dass sie es mir ermöglicht haben, mich auf diese Reise zu begeben. Meiner Frau Franzi danke ich dafür, dass sie es mir ermöglicht hat, diese Reise zu bewältigen. Und meinen beiden Töchtern Matilda und Helene danke ich dafür, dass ich all das jeden Tag mit einem Lächeln tun konnte.

Aachen, im März 2023 *Andreas Kämper*

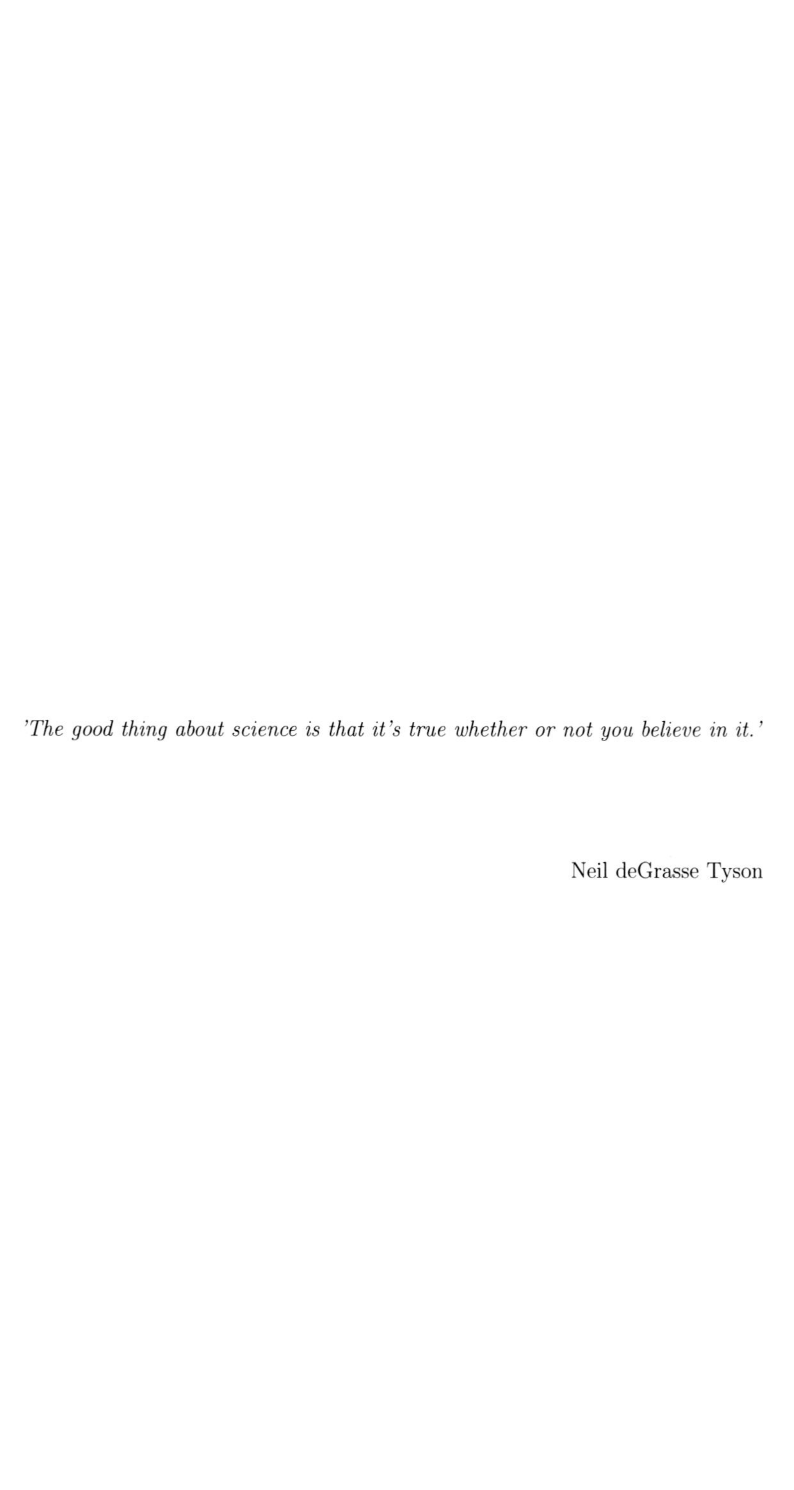

'The good thing about science is that it's true whether or not you believe in it.'

Neil deGrasse Tyson

Contents

I Automated Data-Driven Model Generation of Multi-Energy Systems

II Accelerating Operational Optimization of Multi-Energy Systems

Appendices

List of Figures

List of Tables

Notation

Abbreviations and Acronyms

AIC	Akaike Information Criterion
$\mathrm{AIC_C}$	Corrected Akaike Information Criterion
ANN	artificial neural net
BIC	Bayesian Information Criterion
CHP	combined heat and power
GAMS	General Algebraic Modeling System
GCCM	generalized-convex-combination method
IP	integer programing
LP	linear programing
LSTM	long short-term memory
MILP	mixed-integer linear programing
MINLP	mixed-integer nonlinear programing
MIQP	mixed-integer quadratic programing
NLP	nonlinear programing
ReLU	rectifier linear unit

Latin Symbols

$\boldsymbol{A}$	coefficient matrix
a, A	gradient
$\boldsymbol{b}$	coefficient vector
b, B	intercept
c	cost-based weighting factor
c	specific price
cp	cumulative profit

$CAPEX$	capital expenditure
$\|D\|$	number of data points
d	demand
e	epoch
F_{β}^{micro}	micro-averaged F_{β}-score
h	hyperplane
I	input
i	number of iteration
K	total number of regression parameters
L	number of linear sections
m	Big-M parameter
M	number of dimensions
n	rotational speed
O	output
$OPEX$	operational expenditure
P	power
PoI	period of interest
p	general parameter symbol
Q	amount of heat
q	general parameter symbol
SC	start-up cost
T	number of time steps
t	time step
$\dot{V}$	volumetric flow rate
x	continuous variable
$\boldsymbol{x}$	vector of continuous variables
$\boldsymbol{z}$	vector of all variables

Greek symbols

α	learning rate
γ	binary variable to assign data points to linear sections
δ	binary variable
$\boldsymbol{\delta}$	vector of binary variables
δ	error
Δ	difference

$\boldsymbol{\Delta}$	hinge
ϵ	relative error
ε	sum of squared residuals
η	efficiency
$\boldsymbol{\theta}$	parameter vector
κ	binary variable to denote if the linear section n is used
σ	standard deviation
τ	time step

Subscripts and superscripts

+	upper hinging hyperplane
-	lower hinging hyperplane
'	alternative component
approx	approximation
BE	break even
buy	buying
class	classification
data	measured data
down	component downtime
el	electricity
fore	foresight of Rolling Horizon
full	full time horizon
input	input energy form
max	maximum
min	minimum
nom	nominal operating point
rel	relative
sell	selling
shutdown	component shutdown
sink	energy sink
source	energy source
start	component startup
step	step size of Rolling Horizon
system	multi-energy system
test	test set

train	training set
ub	upper bound
up	component uptime
valid	validation set

Sets

$b \in B$	set of boilers
$chp \in CHP$	set of combined-heat-and-power units
$d \in D$	set of data points
$e \in E$	set of energy forms
$k \in K$	set of subproblems
$n \in N$	set of linear sections
$s \in S$	set of components
$s \in S_e^{\text{input}}$	set of components using energy form e as input
$t \in T$	set of time steps

Kurzfassung

Big Data eröffnet neue Chancen für tiefgreifende Erkenntnisse und die Unterstützung von Entscheidungsprozessen. Um diese Chancen zu nutzen, werden Methoden benötigt, die nutzbares Wissen aus großen Datenmengen ableiten. Solche Methoden können helfen, drängende Herausforderungen in vielen Bereichen zu bewältigen.

Eine drängende Herausforderung für Energiesysteme ist die notwendige Transformation in Richtung Nachhaltigkeit, um den Klimawandel abzuschwächen. Ein wesentlicher Aspekt dieser Herausforderung ist ein dauerhaft optimaler Betrieb der Energiesysteme. Grundsätzlich lässt sich der optimale Betrieb am besten durch mathematische Optimierung ermitteln. Allerdings sind die manuelle Modellerstellung und die Betriebsoptimierung von Energiesystemen zeitaufwändig und können dadurch eine Anwendung mathematischer Optimierung in der Praxis verhindern.

In dieser Arbeit werden Methoden vorgestellt, die anhand von Messdaten automatisch mathematische Modelle von Energiesystemen generieren, um so die zeitaufwändige Modellerstellung deutlich zu verkürzen. Außerdem werden Methoden vorgestellt, die die Betriebsoptimierung von Energiesystemen beschleunigen. Bei der Modellerstellung lösen die vorgestellten datengetriebenen Methoden den Zielkonflikt zwischen Genauigkeit und Berechnungseffizienz des Energiesystemmodells, indem sie jedes Komponentenmodell nach seiner Rolle im Gesamtsystem gewichten. Auf diese Weise erzeugen die Methoden automatisch Energiesystemmodelle, die eine präzise und effiziente Optimierung ermöglichen.

Um die Betriebsoptimierungen von Energiesystemen zu beschleunigen, stellen wir zwei Methoden vor, die die komplexen Betriebsoptimierungsprobleme in kleinere Teilprobleme zerlegen. Dennoch liefern die Methoden qualitativ hochwertige Lösungen. Die erste Methode nutzt Expertenwissen über das spezifische Energiesystem und beschleunigt Betriebsoptimierungen erheblich, während sie eine ausgezeichnete Lösungsqualität beibehält. Die zweite Methode setzt künstliche neuronale Netze ein, um Betriebsoptimierungen in zuverlässig kurzer Zeit bei gleichzeitig hoher Lösungsqualität zu lösen.

Insgesamt ermöglichen die in dieser Arbeit vorgestellten Methoden eine breitere Anwendbarkeit der mathematischen Optimierung für Energiesysteme.

Abstract

Big data raises new opportunities for deep insights and supporting decision-making. To seize these opportunities, methods that derive useful knowledge from large amounts of data are needed. Such methods can help meet urgent challenges in many fields.

An urgent challenge for energy systems is the necessary transformation towards sustainability to mitigate climate change. One crucial aspect of this challenge is a permanent optimal operation of energy systems. In principle, mathematical optimization can best determine the optimal operation of energy systems. However, manual model generation and operational optimization of energy systems are time-consuming and can thus prevent an application of mathematical optimization in practice.

This thesis presents methods that use measured data to automatically generate mathematical models of energy systems to tackle the challenge of time-consuming model generations. Additionally, methods are presented that accelerate the operational optimization of energy systems. Regarding model generation, the presented data-driven methods solve the trade-off between accuracy and computational efficiency of the energy system model by weighting each component model by its role in the overall system. Thereby, the methods automatically generate energy system models that allow for accurate and computationally efficient optimization.

To accelerate the operational optimization of energy systems, we present two methods that decompose the complex operational optimization problem into smaller subproblems. The methods provide high-quality solutions. The first method employs expert knowledge about the individual energy system to significantly accelerate the operational optimization while retaining an excellent solution quality. The second method applies artificial neural nets to solve the operational optimization in a reliably short time while offering a high solution quality.

Overall, the methods presented in this thesis enable a broader application of mathematical optimization for energy systems.

Chapter 1

Introduction

The digital revolution leads us into the era of big data. More than 500 hours of video content are uploaded to YouTube every minute, adding more than 82 years of video content every day (YouTube, 2022); in 2020, WhatsApp delivered more than 100 billion messages every day (Cathcart, 2020). The sheer amount of data offers opportunities for deep insights and supporting decision-making in many fields. However, data alone neither equals knowledge nor wisdom:

> *'We are drowning in information but starved for knowledge.'*
>
> John Naisbitt

To make the best use of the available data, methods are needed to extract relevant patterns in the data, helping to build a bridge between data and knowledge.

One field where such methods are needed is the energy sector, which is undergoing a fundamental and inevitable transformation. The energy sector is a main contributor to greenhouse gas emissions (IPCC, 2013). Greenhouse gas emissions must be massively reduced to limit the increase in the global average temperature well below 2 °C compared to the pre-industrial level (IPCC, 2018). A key to mitigating climate change is the optimal operation of energy systems.

Operational optimization requires energy system models. Today, model generation of energy systems is mainly performed manually and, thus, is time-consuming and expensive. This tedious model generation limits a broad application of operational optimization in engineering practice.

Furthermore, the increasing share of renewable energy resources such as wind and photovoltaics makes the operational optimization of energy systems difficult due to the volatility of these resources. To react to these volatile renewable energy resources on the one hand and to changing energy prices on the other hand, the operational optimization of energy systems must be solved repeatedly and fast. Solving the operational optimization of energy systems fast is often challenging due to the inherent complexity of energy systems leading to computational complexity. This challenge further limits a broader application of operational optimization in practice.

In contrast, inexpensive sensors and data storage have increased the availability of measured data in energy systems. Due to the increasing amount of available measured data, developing methods that use the available data to help generate models and efficiently solve the operational optimization of energy systems is becoming increasingly promising. In particular, for industrial multi-energy systems, the amount of available measured data is increasing due to the implementation of energy management systems (ISO 50001:2018, 2018) and energy audits (DIN EN 16247, 2012).

In this thesis, we propose methods for automated data-driven model generation and methods for accelerating the operational optimization of multi-energy systems. The goal of these methods is to enable a broader application of mathematical optimization in real-world problems.

1.1 Structure of this thesis

In Chapter 2, we[1] review the state-of-the-art in optimization and modeling of multi-energy systems, focusing on operational optimization and data-driven modeling.

The following Chapters 3-6 are structured in two parts. Part I presents methods for automated data-driven model generation of multi-energy systems. The resulting models are suitable for computationally efficient mathematical optimization. Part II presents methods to accelerate the operational optimization of multi-energy systems.

Part I: Automated Data-Driven Model Generation of Multi-Energy Systems

In **Chapter 3**, we present the method AutoMoG for Automated data-driven Model Generation of multi-energy systems. AutoMoG starts with the simplest model of each component of the multi-energy system and adaptively refines one component model per iteration. Importantly, AutoMoG identifies the component model to be refined based on its importance in the overall multi-energy system. Thereby, AutoMoG generates a sufficiently accurate and computationally efficient model of the multi-energy system.

[1]This thesis is written in the *pluralis modestiae*, first to avoid the excessive use of passive voice and second to emphasize that the research behind this thesis is done in collaboration with colleagues, supervised students, and my supervisor. The contribution of the author is pointed out at the beginning of each chapter.

The AutoMoG method from Chapter 3 is limited to multi-energy systems that contain components with one independent variable, e.g., the heat output of a boiler solely depends on its fuel input. In general, a component can have multiple independent variables. For example, the heat output of a combined-heat-and-power plant that consists of a gas turbine and post-firing depends on the heat output of the gas turbine and the gas input of the post-firing. Thus, in **Chapter 4**, we extend the AutoMoG method to AutoMoG 3D. AutoMoG 3D generates models of multi-energy systems that contain components with multiple independent variables.

Part II: Accelerating Operational Optimization of Multi-Energy Systems

In **Chapter 5**, we present the method Adaptive Rolling Horizon to accelerate the operational optimization of multi-energy systems. Adaptive Rolling Horizon decomposes the operational optimization problem into smaller subproblems to reduce the computation time. The size of the subproblems is adapted depending on future energy prices and start-up costs of the components to retain high solution quality. Adaptive Rolling Horizon builds on the widely used Rolling-Horizon method, enabling a broad and quick application.

In **Chapter 6**, we present a method to solve the operational optimization of multi-energy systems using artificial neural nets. The method decomposes the operational optimization into single-time-step optimizations. The single-time-step optimizations incorporate predictions of key variables from the artificial neural nets that are trained on long-term operational optimizations. Thereby, the method provides a feasible solution for the operational optimization of multi-energy systems in a reliably short time.

Finally, in **Chapter 7**, the thesis is summarized, and conclusions are drawn. Furthermore, perspectives for future research in the field of data-driven multi-energy system modeling and optimization are discussed.

Chapter 2

State-of-the-art in Optimization and Modeling of Multi-Energy Systems

This chapter provides an overview of the state-of-the-art in optimization and modeling of multi-energy systems. In Section 2.1, we present a general introduction to multi-energy system optimization. In particular, we present an introduction to the optimal operation of multi-energy systems and discuss the requirements for their practical application. In Section 2.2, we introduce the state-of-the-art in modeling multi-energy systems with a focus on data-driven modeling. In Section 2.3, we review methods to accelerate the operational optimization of multi-energy systems. Finally, in Section 2.4, we state the contribution of this thesis.

Kämper, A., Leenders, L., Bahl, B., and Bardow, A. (2021c). AutoMoG: Automated data-driven Model Generation of multi-energy systems using piecewise-linear regression. *Computers & Chemical Engineering*, 145, 107162.

Kämper, A., Holtwerth, A., Leenders, L., and Bardow, A. (2021b). AutoMoG 3D: Automated Data-Driven Model Generation of Multi-Energy Systems Using Hinging Hyperplanes. *Frontiers in Energy Research*, 9, 719658.

Kämper, A., Geers, P., Leenders, L., and Bardow, A. (2021a). Adaptive Rolling Horizon for operational optimization of multi-energy systems. In *Proceedings of the ECOS 2021 - 34th International Conference on Efficiency, Cost, Optimization, Simulation and Environmental Impact of Energy Systems (ECOS 2021)*.

Kämper, A., Delorme, R., Leenders, L., and Bardow, A. (2023). Boosting Operational Optimization of Multi-Energy Systems by Artificial Neural Nets. *Computers & Chemical Engineering*, 173, 108208.

Contribution report: Principal author, Conceptualization, Methodology, Software, Validation, Formal Analysis, Investigation, Data Curation, Writing - Original Draft, Visualization, Project administration.

2.1 Optimization of multi-energy systems

Energy systems comprise all components related to the conversion, delivery, and use of energy (IPCC, 2014). The scale of energy systems reaches from small, decentralized units supplying households to large-scale energy systems supplying a whole country or even multiple countries.

In this thesis, we consider energy systems that supply industrial sites. These energy systems are usually located on-site. A major advantage of on-site energy systems is avoidance of final energy losses due to long transport (Voll, 2014). Here, we focus on multi-energy systems (Mancarella, 2014) that consist of many components. A component is a physical unit or a subsection of the multi-energy system that stores energy or converts primary or secondary energy to useful forms of energy (Frangopoulos, 2018). Multi-energy systems supply multiple energy demands such as heating, cooling, electricity, and steam. Multi-energy systems are regarded as a key element of future sustainable energy systems since they can efficiently integrate the conversion of several energy inputs and outputs (Mancarella et al., 2016; Guelpa et al., 2019; Moretti et al., 2020). However, the integration of many components in a multi-energy system results in an inherently high level of complexity. For this reason, the optimal design and operation of multi-energy systems are best addressed by mathematical optimization (Mancarella, 2014; Andiappan, 2017; Thie et al., 2020).

A generic mathematical optimization problem is stated in Equation (2.1). In an optimization problem, an objective function f is minimized by determining values for continuous decision variables $\mathbf{x}$ and discrete decision variables $\boldsymbol{\delta}$. The values of the decision variables $\mathbf{x}$ and $\boldsymbol{\delta}$ are subject to a set of inequality constraints g and a set of equality constraints h.

$$
\begin{aligned}
\min_{\mathbf{x},\boldsymbol{\delta}} \quad & f(\mathbf{x},\boldsymbol{\delta}) \\
\text{s.t.} \quad & g(\mathbf{x},\boldsymbol{\delta}) \geq 0, \\
& h(\mathbf{x},\boldsymbol{\delta}) = 0, \\
& \mathbf{x} \in \mathbb{R}^{\mathrm{m}}, \boldsymbol{\delta} \in \{0,1\}^{\mathrm{n}}.
\end{aligned} \tag{2.1}
$$

The sets of inequality constraints g and equality constraints h describe the relations between the decision variables $\mathbf{x}$ and $\boldsymbol{\delta}$ and limit their values. Optimization problems can be classified depending on whether continuous decision variables $\mathbf{x}$ and/or discrete decision variables $\boldsymbol{\delta}$ occur, and whether the functions f, g, and h are linear or nonlinear: linear programming (LP), nonlinear programming (NLP), integer programming

(IP), mixed-integer linear programming (MILP), and mixed-integer nonlinear programming (MINLP).

In multi-energy system models, typical inequality constraints are limits of component inputs or outputs, and typical equality constraints are energy balances. For optimally designing multi-energy systems, a common objective is to minimize the sum of annualized capital expenditures $CAPEX$ and operational expenditures $OPEX$. For optimally operating multi-energy systems, a common objective is to minimize the operational expenditures $OPEX$. Multi-energy system models typically contain discrete decision variables, e.g., the existence and the on/off status of all components, and continuous variables, e.g., components' loads.

To find optimal multi-energy system designs, the optimization models have to contain the nonlinear investment-cost curves to represent the components' $CAPEX$. Additionally, the optimization models have to contain nonlinear input-output relationships of all components to represent the part-load performance. Thus, in general, mathematical optimization of multi-energy system designs leads to mixed-integer nonlinear programming problems (MINLP) (Bruno et al., 1998). In an operational optimization, the design variables are fixed. However, the optimization models still include the usually nonlinear input-output relationship of each component, which leads to MINLP problems.

MINLPs are challenging to solve to global optimality (Mitsos et al., 2018). Commonly, nonlinearities are therefore approximated by piecewise-affine models leading to Mixed-Integer Linear Programs (MILPs) (Gao et al., 2018; Voll et al., 2013; Misener et al., 2009). MILPs enable finding the global optimum efficiently with established solvers (Grossmann, 2012; Floudas, 1995). For this reason, the operational optimization of multi-energy systems is often modeled as an MILP (Elia and Floudas, 2014; Kantor et al., 2020). However, the computation time of an MILP highly varies for each concrete instance and is challenging to predict (Yang and Wu, 2021). In fact, the complexity of an MILP operational optimization of multi-energy systems is at least weakly NP-hard even for a single time step (unless P=NP) (Goderbauer et al., 2019).

Additionally, an operational optimization is typically time-coupled, e.g., by storage, ramping constraints, or start-up costs. Thus, future energy demands and prices affect the optimal decisions of the present. In general, the more time steps in the operational optimization, the more information about future energy demands and prices, the better the decisions for today. However, time-coupling prevents a simple decomposition of the operational optimization problem to single time steps (Cao et al., 2019). The problem size increases linearly with the number of time steps, and the computational effort to solve the problem increases exponentially with the problem size (Kallrath,

2000). Thus, the computation time increases drastically with the number of considered time steps such that time and computational limits are quickly reached.

This challenging operational optimization problem has to be solved repeatedly in practice to reflect changes in demands, resource availability, or prices (Wang et al., 2015). Thus, the computation time is a practical concern for operational optimization of multi-energy systems, particularly when participating in electricity markets due to short market-clearing windows (Yang and Wu, 2021). The optimal operation relies on energy-price predictions such that the quality of the predictions determines the solution quality of the operational optimization. Those energy-price predictions become more accurate the closer the time of the prediction is to the predicted time. However, the closer the predictions are made to the predicted time, the less time is left for operational optimization. Hence, the operational optimization problem of multi-energy systems needs to be solved fast.

As a result, solving MILPs for operational optimization of multi-energy systems in a reliably short time and with good solution quality is desired. For this purpose, two main challenges have to be addressed: The first main challenge concerns model generation. Model generation determines the accuracy and computational efficiency of multi-energy system models. Thus, we have to generate multi-energy system models that are sufficiently accurate to identify the correct operating decisions that minimize $OPEX$. However, the multi-energy system models should be as simple as possible to be computationally efficient. For this reason, we review methods to model multi-energy systems in Section 2.2. The second main challenge concerns the operational optimization itself. The complexity of the multi-energy system model is fixed after modeling. However, the solution method for the operational optimization problem still has a significant impact on the computation time. For this reason, we review solution methods that focus on accelerating the operational optimization of multi-energy systems in Section 2.3.

2.2 Modeling of multi-energy systems

When modeling multi-energy systems, we should be aware that all models are wrong, but some are useful (Box, 1979). Modeling always needs to consider the model's purpose since a model always balances accuracy and computational efficiency. For generating meaningful insights, the multi-energy system models have to be sufficiently accurate (Welsch et al., 2014). However, for operational optimization, the models also have to be computationally efficient to respect time limits. Generating multi-energy system models that are both accurate and computationally efficient is challenging

(Mitsos et al., 2018). Consequently, a model should be as simple as possible and as complex as necessary to fulfill its purpose properly (DeCarolis et al., 2017).

Today, models of multi-energy systems are commonly generated manually. Thus, model generation requires high effort and is time-consuming (Bonvin et al., 2016).

In the following sections, we first present fundamental modeling approaches. Afterwards, we focus on data-driven modeling.

2.2.1 Modeling approaches

In general, three approaches can be followed to generate models: first-principles modeling (also referred to as white-box modeling), data driven modeling (also referred to as black-box modeling), and a hybrid approach that combines first-principles and data-driven modeling (also referred to as grey-box modeling) (Demirhan et al., 2019). First-principles models are derived from theory with the aim to represent the real physical behavior of a system or component (Smolin et al., 2019). For this reason, first-principles models are easily interpretable. However, solving full first-principles models often is computationally demanding (McBride and Sundmacher, 2019). Furthermore, the physical behavior of the system is frequently partly unknown, which prevents full first-principles modeling.

Data-driven models are derived from data with the aim to represent the input-output relationship of a system (McBride and Sundmacher, 2019). Measured data is increasingly available in multi-energy systems, in particular, due to the implementation of energy management systems according to ISO 50001:2018 (2018). Thus, data-driven model generation for multi-energy systems becomes increasingly promising. However, data-driven models are typically hard to interpret.

Thus, in this thesis, we generate hybrid models of multi-energy systems since they combine the advantages of first-principles and data-driven models (Li et al., 2021). The generated multi-energy system models contain data-driven component models and first-principle constraints, e.g., energy balances. In the next section, we review methods for data-driven modeling of multi-energy system components.

2.2.2 Data-driven modeling

In a multi-energy system, measured input and output data can be used to generate a data-driven model of each component. In general, the measured data can be used to correlate the input-output relationships of the components through regressions. One regression problem has to be solved per component. The regression approximates a functional relationship between independent input variables and output variables from a given data set (Yang et al., 2016). The data-driven component models are then joined together and expanded by additional constraints, e.g., energy balances, to obtain one model of the multi-energy system. As aforementioned, the regressions should yield computationally efficient models. At the same time, the models have to be sufficiently accurate.

For regression, many approaches are available such as linear regression (Weisberg, 2005), kriging (Chilès and Delfiner, 2012), support-vector regression (Awad and Khanna, 2015; Smola and Schölkopf, 2004), or neural networks (Huang et al., 2010).

To generate accurate and computationally efficient models, Cozad et al. (2014) and Wilson and Sahinidis (2017) present a framework for automated learning of algebraic models (ALAMO). ALAMO provides black-box models from data obtained by simulation or experiments. Cozad et al. (2014) use the Corrected Akaike Information Criterion AIC_{C} (Hurvich and Tsai, 1993) to find simple models with sufficient accuracy. Information criteria select the most suitable model for a data set by balancing model accuracy and model complexity. Widely known information criteria are, e.g., the Akaike Information Criterion AIC (Akaike, 1974) or the Bayesian Information Criterion BIC (Stoica and Selén, 2004). Since the Akaike Information Criterion AIC shows a bias for small sample sizes, Hurvich and Tsai (1993) developed the Corrected Akaike Information Criterion AIC_{C} that is also suitable for small sample sizes.

The resulting models from ALAMO are generally nonlinear. If these nonlinear models are used for operational optimization, the problem will usually be a Mixed-Integer Nonlinear Program (MINLP). In general, MINLP problems are harder to solve, and a global optimum cannot be guaranteed (Mitsos et al., 2018).

To obtain an MILP for the operational optimization, the component models have to be piecewise affine. The more affine regions, the better the model can adapt to the measured data, but the more complex the resulting model. Thus, the objective of the model generation is to use as few affine regions per component as possible but as many as necessary for sufficient accuracy. In general, the generation of accurate and computationally efficient piecewise-affine models yields complex MINLPs itself and is, therefore, an active field of research. Zhang et al. (2016) propose a data-driven algorithm to generate surrogate models of process systems. The generated surrogate models are piecewise affine in convex regions and, thus, can be used in MILPs. Yang et al. (2016) and Gkioulekas and Papageorgiou (2018) provide a mathematical programming approach for piecewise-affine regression. The piecewise-affine regression models are obtained by solving MILP regression problems. MILP regression problems minimize the distance between data and model to retain linearity. Additionally, Kong and Maravelias (2020) and Rebennack and Krasko (2020) formulate MILP approaches for continuous piecewise-affine regression. These MILP approaches derive univariate piecewise-affine models from measured data. As the provided models are piecewise affine, they can be used in MILPs.

The reviewed methods can solve the piecewise-affine regression problem for any component in a multi-energy system. However, the reviewed methods are restricted to multi-energy systems that contain components with one independent variable, e.g., the heat output of a boiler solely depends on its fuel input. In general, a component's input-output relationship depends on multiple independent variables, e.g., the power consumption of a pump depends on its rotational speed and volumetric flow rate. Another typical component in energy systems is a combined-heat-and-power (CHP) plant. The heat output of a CHP plant consisting of a gas turbine and post-firing depends on the heat output of the gas turbine and the gas input of the post-firing (Bischi et al., 2014). Deriving these input-output relationships from measured data leads to multidimensional piecewise-affine regression problems.

Various approaches generate multidimensional piecewise-affine models that are suitable for MILP optimization. Fischetti and Jo (2018) and Grimstad and Andersson (2019) show that deep neural networks with rectifier linear units as activation functions can be formulated as MILP models. Thus, the deep neural networks can approximate a nonlinear model with arbitrary accuracy and then be embedded in a subsequent MILP optimization (Schweidtmann and Mitsos, 2019; Katz et al., 2020). However, embedding deep neural networks in optimization problems that remain computationally efficient is challenging and, thus, an active field of research (Anderson et al., 2020; Yang et al., 2021).

Obermeier et al. (2021) propose two approaches to generate multidimensional piecewise-affine models from measured data. Both approaches create a mesh using all data points with Delauney triangulation (Barber et al., 1996). The first approach called Iterative Mesh Reduction iteratively reduces the complexity of the created mesh by contracting the edges of this mesh. The second approach called Iterative Mesh Refinement chooses one affine region of the created mesh to represent all data points. Then, Iterative Mesh Refinement iteratively increases the affine regions to represent all data points until a predefined accuracy is reached. Furthermore, Kazda and Li (2021) propose a method for multidimensional piecewise-affine fitting using the difference-of-convex representation. The method aims to generate a model with predefined accuracy while keeping the number of affine regions low. The approaches of Obermeier et al. (2021) and Kazda and Li (2021) are designed to transform a well-defined functional relationship or handle noiseless data and not to handle noisy data. However, measured data obtained in real-world applications are typically noisy.

Hinging hyperplanes (Breiman, 1993) have been shown to handle well-defined functional relationships as well as noisy measured data (Roll et al., 2004). Hinging-hyperplane-tree regression (Ernst, 1998) is based on the hinge-finding algorithm proposed by Breiman (1993) and can solve multidimensional piecewise-affine regression problems. The original hinge-finding algorithm (Breiman, 1993) suffers from convergence problems. To overcome this drawback, an improved hinge-finding algorithm has been developed (Pucar and Sjöberg, 1998). Roll et al. (2004) formulate an MILP approach using hinging hyperplanes to provide the global optimum of multidimensional piecewise-affine regression problems at the price of increased computational effort.

Altogether, numerous approaches exist to solve unidimensional and multidimensional piecewise-affine regression problems. These approaches can be used to generate data-driven models of arbitrary multi-energy system components for MILP operational optimization. However, while these approaches overcome the MINLP problem, they do not reflect the complex structure of multi-energy systems. A multi-energy system model that is sufficiently accurate does not necessarily require an accurate model of all components in the multi-energy system. Thus, independently modeling each component of a multi-energy system may lead to an overall model of the multi-energy system that is unnecessarily complicated and, thus, computationally inefficient (Qin et al., 2021).

2.3 Solution methods for operational optimization of multi-energy systems

The complexity of the component models and the multi-energy system model itself is fixed after modeling. However, the method to solve an MILP problem still has a significant impact on the computation time and is therefore important for operational optimization. An easy and reasonable method for solving an MILP problem is applying a state-of-the-art solver. As aforementioned, state-of-the-art solvers like Gurobi (Gurobi Optimization, 2020), CPLEX (IBM Corporation, 2017), or SCIP (Gleixner et al., 2018) can efficiently solve MILP problems. The solver performance has made huge progress in the past (Bixby, 2020; Achterberg and Wunderling, 2013) and enables solving optimization problems of complex systems in the first place. However, optimization problems of complex multi-energy systems are still computationally demanding (Bahl et al., 2018b). For this reason, applying a state-of-the-art solver is not sufficient to respect time limits of operational optimization. Thus, operational optimization needs to be accelerated. Due to typically strict time limits of operational optimization, solution methods aim to provide high-quality solutions in a *reliably short time.* We use the term *reliably short time* to refer to a time that, regardless of the specific problem instance, solves the optimization model in a predictable time short enough to meet the strict time limits of operational optimization.

In the following, we review decomposition methods and machine-learning approaches that reduce computational effort and, thereby, accelerate operational optimization.

2.3.1 Decomposition methods

A common solution approach to accelerate optimization problems is decomposition. For example, Benders' decomposition (Benders, 1962) can be used to solve problems with time-coupling variables. A typical time-coupling variable in multi-energy systems is a storage level (Baumgärtner et al., 2020). Nasrolahpour et al. (2016) use Benders' decomposition in a stochastic bi-level optimization problem to optimally size and operate energy storage. Rahmaniani et al. (2017) list numerous approaches that improve the performance of Benders' decomposition and expand its applicability. However, time-consuming iterations, ineffective initial iterations, and slow convergence at the end of the algorithm limit the acceleration by Benders' decomposition (Rahmaniani et al., 2017). Furthermore, formulating and implementing Benders' decomposition remains complex.

Lagrangian relaxation is able to solve problems with time-coupling constraints (Rong et al., 2008). A time-coupling constraint is, e.g., an overall emission limit that must not be exceeded over a specific period, typically a year. The convergence of Lagrangian relaxation highly depends on the chosen Lagrangian multipliers. However, there is no straightforward approach for selecting Lagrangian multipliers, hindering an easy application to a wide range of problems (Conejo et al., 2006).

Dantzig-Wolfe decomposition can also handle time-coupling constraints (Dantzig and Wolfe, 1960). Yokoyama and Ito (1996) applied Dantzig-Wolfe decomposition to an MILP operational optimization of thermal storage systems. To apply Dantzig-Wolfe decomposition, knowledge about the specific problem structure is necessary. To overcome this drawback, Bergner et al. (2015) automate the problem reformulation to apply Dantzig-Wolfe decomposition. However, the computation times remain high for many problems. In fact, it is often questionable if applying Dantzig-Wolfe decomposition accelerates the solving process at all. For this reason, Kruber et al. (2017)

train a machine-learning model to decide whether or not applying Dantzig-Wolfe decomposition reduces the computation time.

To solve problems with time-coupling constraints and variables, Wang et al. (2016) combine Benders' decomposition and Lagrangian relaxation. Still, the major drawbacks of the Benders' decomposition and Lagrangian relaxation remain present. For this reason, Baumgärtner et al. (2020) present the method DeLoop. DeLoop iteratively decomposes long-term operational optimization of one year into smaller subproblems and recombines their solutions to provide a feasible solution for the long-term operational optimization with known quality. The method is significantly faster than CPLEX (up to 32 times faster) and solves the operational optimization of their case study in nearly 1.5 h. However, this computation time is not sufficiently short for repeated short-term operational optimization, and the computation time would increase for more complex multi-energy systems. Additionally, the method requires accurate forecasts for demands and prices of one year to provide a viable solution.

The decomposition methods mentioned above can accelerate operational optimization with known solution quality. However, the acceleration still does not reach suitable computation times for real-world problems. Additionally, applying these methods requires high effort.

An easy-to-implement approach to reduce computation time is the Rolling-Horizon method (Cao et al., 2019). For this reason, Rolling Horizon is often used in practice to reduce computation time (Bischi et al., 2017; Shu et al., 2019; Shin and Zavala, 2020) or to consider uncertainties (Gupta et al., 2016; Wang et al., 2015; Kopanos and Pistikopoulos, 2014). Rolling Horizon heuristically decomposes the time axis of the original optimization problem into smaller subproblems that consist of a step size T^{step} and a foresight T^{fore} (Figure 2.1).

The subproblems are solved successively. Only the solutions for the step size T^{step} are merged into the Rolling-Horizon solution of the original operational optimization. The subproblems are extended by a foresight T^{fore} such that they contain more information about the future, e.g., energy demands and prices. The solutions for the foresight T^{fore} are neglected but improve the decisions in the step size T^{step}. Altogether, Rolling Horizon is easily applicable to many problems and can heavily reduce the computation time (Marquant et al., 2015). Applying Rolling Horizon requires the selection of values for the step size and foresight, and the computation time and solution quality strongly depend on the chosen values for the step size and the foresight (Marquant et al., 2015). We observe the following main trends: The larger the step size, the fewer subproblems to solve, but the longer the computation time of each subproblem. Thus, the choice of the step size results mainly in a trade-off in com-

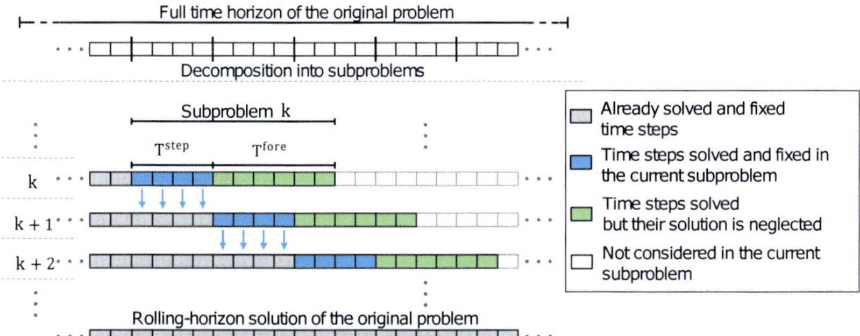

Figure 2.1: Procedure of the Rolling-Horizon method. The original problem is decomposed into smaller subproblems k. The subproblems are solved successively. The solutions for the step size T^{step} are merged into the Rolling-Horizon solution of the original problem. The solution for the foresight T^{fore} is neglected but can influence the decisions in the step size T^{step}.

putation time. For the foresight, an increase improves the solution quality but leads to a longer computation time of each subproblem. Thus, the choice of the foresight results in a trade-off between short computation time and high solution quality. Thus, choosing values for the step size and the foresight is not straightforward. There is no approach to select values for the step size and the foresight such that computation time is heavily reduced while solution quality remains high.

2.3.2 Combining machine learning and mathematical optimization

Recently, many works have combined machine-learning techniques and mathematical optimization to decrease the computational effort for solving optimization problems (Ruan et al., 2020). Machine learning can be viewed as a set of methods that can automatically detect patterns in data and then use the patterns to predict future data or perform other kinds of decision-making (Murphy, 2012). In particular, artificial neural networks (ANNs) have gained much attention due to their ability to approximate arbitrary functions (Hornik et al., 1989). For an introduction and an overview of machine learning, the interested reader is referred to Bishop (2016), Murphy (2012), and Goodfellow et al. (2016).

There are three principal approaches to help solve MILPs using machine learning (Bengio et al., 2021): end-to-end learning, learning to configure optimization algorithms, and machine learning alongside optimization algorithms. In end-to-end learning, a machine-learning model replaces the optimization model and provides a solution directly. For example, Vinyals et al. (2015) introduce a pointer network to tackle the traveling-salesman problem. The pointer network is trained using supervised learning with precomputed traveling-salesman-problem solutions as targets. Once trained, the machine-learning models provide a solution fast and in a reliably short time. However, in end-to-end learning, the machine-learning model learns a heuristic. Therefore, the models generally do not guarantee feasibility (Bengio et al., 2021).

The second approach is learning to configure solution algorithms. Here, machine learning is used to augment an optimization algorithm with valuable information (Nair et al., 2021). In a broad sense, machine learning provides a parametrization of the solution algorithm (Bengio et al., 2021). For example, machine-learning techniques learn to choose values for the step size in gradient descent methods of MILP solvers to speed up the optimization algorithm (Bengio et al., 2021). In this context, Kruber et al. (2017) apply machine learning to decide whether or not applying Dantzig-Wolfe decomposition reduces the computation time. Bonami et al. (2018) use machine-learning models to decide if linearizing mixed-integer quadratic programming problems leads to faster solutions. Masti and Bemporad (2019) use a neural network to provide binary warm starts for solving mixed-integer quadratic programs. The reviewed approaches that learn to configure solution algorithms are widely applicable. While the approaches can often reduce the computation time relatively, none of the approaches reduces the computation time to a reliably short time. Furthermore, the approaches are hard to implement and require the solution algorithm to exchange detailed information while solving an MILP.

The third approach is machine learning alongside optimization algorithms. Here, the MILP optimization algorithm repeatedly queries the same machine-learning model to make decisions (Zarpellon et al., 2021). Learning branching policies for branch-and-bound algorithms to solve MILPs is a good example of this approach. Selecting variables and nodes to branch on is crucial in branch-and-bound algorithms. The selection is often only based on heuristics or too slow. Machine learning is a natural candidate to tackle this problem (Lodi and Zarpellon, 2017). Alvarez et al. (2017) use Extremely randomized trees (Geurts et al., 2006) to learn the branching decisions from the branching strategy Strong Branching. Strong Branching provides good branching decisions but is computationally expensive. It tentatively branches on many candidate variables and then selects the branch on the variable that promises the best lower-bound improvement. Gasse et al. (2019) also learn from Strong Branching but use a neural net to decide which branching node to explore next. The approaches in repeated exchange with solution algorithms are also widely applicable but show the same weaknesses as the second approach mentioned above.

For operational optimization, literature on combining machine learning and optimization is limited (Yang and Wu, 2021). Xavier et al. (2020) present a method to solve large-scale security-constrained unit commitment problems by machine learning. For this purpose, three machine-learning models are used: The first model predicts which constraints are necessary and which can be omitted. The second model predicts a partial solution that works as a warm start. Finally, the third model identifies a solution subspace where the optimal solution is likely to lie. The computation time is reduced by a factor of up to 10. Although this relative speed-up is significant, the method does not solve the operational optimization in a reliably short time.

Considering multi-energy systems in particular, only a few studies use approaches that combine machine learning and optimization (Alabi et al., 2022a). Taheri et al. (2021) used a deep recurrent neural net for the long-term planning and design of multi-energy systems. The approach uses Long-Short-Term-Memory (LSTM) layers (Hochreiter and Schmidhuber, 1997) to predict heat and electricity demands in hourly resolution. Subsequently, gaussian process regression is used to provide a solution without explicitly solving an optimization problem. Kong et al. (2020) use a generative adversarial net to generate multi-energy load scenarios. Based on these scenarios, a two-stage robust stochastic optimization solves the scheduling problem of a multi-energy virtual power plant. Similarly, Alabi et al. (2022b) combine deep learning for prediction and optimization methods for optimal scheduling of multi-energy systems. Zhou et al. (2019) use a LSTM model to predict the operation of a hydrogen-penetrated energy system. The LSTM model learns from pre-solved MILP solutions. However, the computation time of the operational optimization is not examined. Additionally, the optimization problem is relatively small (11 variables over 24 h).

The above-mentioned approaches show that combining machine learning and optimization is possible and promising for multi-energy systems. However, none of the approaches uses a combined approach that tackles the problem of high and unreliable computation times in operational optimizations of multi-energy systems. In fact, it is observed that applying a combination of machine learning and optimization to the optimal decision-making on multi-energy systems is at an early stage that requires further research (Alabi et al., 2022a). In particular, a solution method is still missing that is applicable to large-scale MILP multi-energy system models and provides feasible solutions for operational optimization in a reliably short time.

2.4 Contribution of this thesis

Two major tasks have to be fulfilled to solve the operational optimization of multi-energy systems: the first task is generating a sufficiently accurate model, and the second task is using the model in an operational optimization to find a high-quality solution in a short time. These two tasks are commonly time-consuming and thus expensive. The model generation is mainly done manually in practice. To reduce this high effort, methods are desired that automate the model generation from measured data. Operational optimizations require high computation times even when using strong, commercial solvers due to the complexity of multi-energy systems. Thus, solutions methods are required that are easily applicable to multi-energy system models and provide high-quality solutions fast.

The literature review in Section 2.2 reveals that existing modeling methods can derive MILP models from measured data. However, they do not reflect the complex structure of multi-energy systems since they model each component independently. Thus, methods that automatically generate accurate and computationally efficient MILP models of multi-energy systems are missing.

Section 2.3 shows that methods exist that can accelerate MILP operational optimization. However, no easy-to-apply method substantially accelerates the operational optimization while retaining high solution quality. Consequently, none of the methods provides high-quality solutions for the operational optimization of multi-energy systems in a reliably short time.

This thesis closes the identified gaps by providing methods that cover the complete process from measured data to optimal operation of multi-energy systems (Figure 2.2). The thesis is structured in two parts: Part I is dedicated to data-driven modeling. Part II is dedicated to accelerating the operational optimization of given MILP multi-energy system models.

Part I: Automated data-driven model generation of multi-energy systems

Chapter 3: AutoMoG. The data-driven model generation is a challenging problem for multi-energy systems that typically consist of many components. For each component in the system, the modeler has to solve a regression problem to represent its input-output relationship. Typically, the modeler generates the model of each component independently. However, this independent modeling may lead to models that are unnecessarily complicated and, thus, computationally inefficient. Thus, we propose

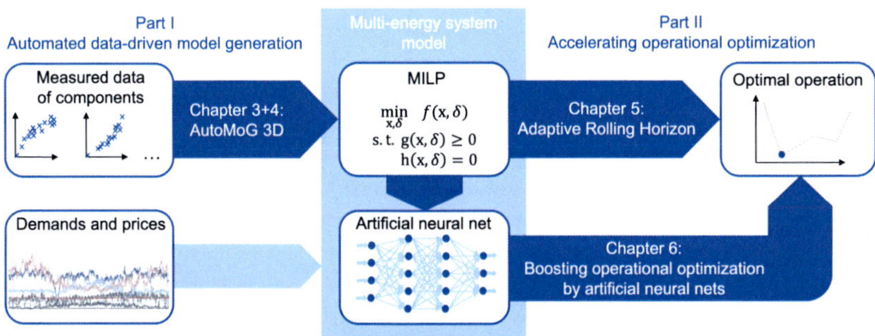

Figure 2.2: Contribution of this thesis: methods for automated data-driven model generation and operational optimization enabling the efficient performance of the full process from measured data to optimal operation of multi-energy systems.

the method AutoMoG for Automated data-driven Model Generation of multi-energy systems. In contrast to common practice, AutoMoG does not model each component independently, which may lead to an unnecessarily complicated model of the overall multi-energy system. Instead, AutoMoG balances the errors caused by the model of each component in the overall multi-energy system model. Thereby, each component is modeled as simply as possible and as accurately as necessary to obtain an accurate and computationally efficient model of the multi-energy system.

Chapter 4: AutoMoG 3D. The method AutoMoG is limited to multi-energy systems that contain components with one independent variable, e.g., the heat output of a boiler solely depends on its fuel input. In general, components of multi-energy systems have multiple independent variables, e.g., the power consumption of a pump depends on its rotational speed and volumetric flow rate. Thus, we extend AutoMoG to AutoMoG 3D to automatically generate MILP optimization models of multi-energy systems that contain components with multiple independent variables. AutoMoG 3D builds on the advantages of the AutoMoG method (see Chapter 3), i.e., AutoMoG 3D considers the importance of the components in the context of the overall multi-energy system.

Part II: Accelerating operational optimization of multi-energy systems

Chapter 5: Adaptive Rolling Horizon. Operational optimization of multi-energy systems needs to be accelerated to respect the typically strict time limits. Thus, solutions methods are required that are easily applicable to multi-energy system models and provide high-quality solutions fast. Typical Rolling Horizon is easy to apply to a wide range of problems. Applying Rolling Horizon requires the selection of values for the two parameters, step size and foresight. Selecting these values is not straightforward. The computation time and solution quality strongly depend on the number of time steps in the step size and the foresight (Marquant et al., 2015). There is no approach to select values for the step size and the foresight such that computation time is heavily reduced while solution quality remains high. To close this gap, we propose the Adaptive-Rolling-Horizon approach. The approach uses a fixed step size T^{step} and adaptively determines the foresight T_k^{fore} for each subproblem k by employing expert knowledge about the multi-energy system.

Chapter 6: Boosting operational optimization by artificial neural networks. The Adaptive-Rolling-Horizon approach substantially accelerates the operational optimization of multi-energy systems and reaches a high solution quality. However, the Adaptive Rolling Horizon does not solve operational optimizations in a reliably short time. Additionally, the approach is not suitable for multi-energy systems containing energy storage units. Thus, we propose a new method that combines machine learning and combinatorial optimization to solve operational optimization problems of large-scale multi-energy systems. By decomposing the original optimization problem into single time steps, the method reaches reliably short computation times. Still, artificial neural nets allow integrating long-term effects in the single-time-step optimizations. In contrast to the Adaptive Rolling Horizon, the proposed decomposition method is suitable for multi-energy systems containing energy storage units.

Concluding, this thesis proposes methods that use data of multi-energy systems to automatically generate mathematical models and provide high-quality solutions for operational optimization in a short time. The proposed methods cover the complete process from measured data to optimal operation of a multi-energy system.

Part I

Automated Data-Driven Model Generation of Multi-Energy Systems

Chapter 3

AutoMoG: Automated Data-Driven Model Generation using Piecewise-Linear Regression

In this chapter, we propose the method AutoMoG for Automated data-driven Model Generation of multi-energy systems. AutoMoG copes with the first major task that we identified in Chapter 2.4: generating sufficiently accurate and computationally efficient multi-energy system models with measured data. AutoMoG solves the data-driven model generation problem by piecewise-linear regression[1] while balancing the errors caused by each component's model in the overall multi-energy system model. Thereby, AutoMoG automatically generates multi-energy system models that are accurate and computationally efficient. To enable a straightforward application of AutoMoG, we released a Python version of the code open-source (Kämper et al., 2021d).

The chapter is structured as follows: In Section 3.1, we formulate the data-driven model generation problem for multi-energy systems. In Section 3.2, we describe the proposed method AutoMoG. In Section 3.3, we apply AutoMoG to a case study for a multi-energy system from literature. In Section 3.4, we conclude with the key findings.

Major parts of this chapter are reproduced by permission of Elsevier from:

Kämper, A., Leenders, L., Bahl, B., and Bardow, A. (2021c). AutoMoG: Automated data-driven Model Generation of multi-energy systems using piecewise linear regression. *Computers & Chemical Engineering*, 145, 107162.

Contribution report: Principal author, Conceptualization, Methodology, Software, Validation, Formal Analysis, Investigation, Data Curation, Writing - Original Draft, Visualization, Project administration.

[1]In this chapter, we use the term *piecewise-linear regression* since it is commonly used in the literature. However, we use *piecewise-affine* to describe the resulting models since it is the correct mathematical term. Piecewise-affine functions do not include the origin in general.

3.1 Data-driven model generation for multi-energy systems

The data-driven model generation problem for multi-energy systems shall provide a sufficiently accurate and computationally efficient MILP model of the multi-energy system. As mentioned in Chapter 2.1, MILP models enable finding the global optimum efficiently with established solvers. In the provided MILP model, the input-output relationship of each component $s \in S$ in the multi-energy system has to be represented by a piecewise-affine model.

In general, the functional relationship between input I_s (e.g., gas or electricity) and output O_s (e.g., heating or cooling) of a component s (e.g., boiler or compression chiller) is nonlinear. For MILP optimization models, nonlinear functional relationships are approximated by piecewise-affine functions $I_s^{\text{model}}(O_{s,n})$. Here, we choose to model the input I_s^{model} as a linear function of the output $O_{s,n}$, because we can easily convert the input to operating cost. This conversion is crucial for the AutoMoG method; more details are given in Section 3.2.1. However, AutoMoG can be easily adapted to model the output as a function of the input.

$$I_s^{\text{model}} = \sum_{n \in N_s} \left(\hat{O}_{s,n} a_{s,n} + \hat{\delta}_{s,n} b_{s,n} \right), \qquad \forall s \in S, \tag{3.1}$$

$$\hat{O}_{s,n} \leq \hat{\delta}_{s,n} o_{s,n}^{\text{ub}}, \qquad \forall s \in S, n \in N_s, \tag{3.2}$$

$$\hat{O}_{s,n} \geq \hat{\delta}_{s,n} o_{s,n-1}^{\text{ub}}, \qquad \forall s \in S, n \in N_s \setminus \{0\}, \tag{3.3}$$

$$O_s = \sum_{n \in N_s} \hat{O}_{s,n}, \qquad \forall s \in S, \tag{3.4}$$

$$\delta_s = \sum_{n \in N_s} \hat{\delta}_{s,n}, \qquad \forall s \in S, n \in N_s, \tag{3.5}$$

with I_s^{model} being the modeled input of component s and N_s being the set of piecewise-affine sections n of component s. The parameter $a_{s,n}$ denotes the slope of linear section n, and the parameter $b_{s,n}$ denotes the intercept of linear section n. The binary variable δ_s denotes the on/off state of the component s. The binary variable $\hat{\delta}_{s,n}$ is equal to one if and only if the output O_s lies between the upper bound $o_{s,n}^{\text{ub}}$ of the linear section n and the upper bound $o_{s,n-1}^{\text{ub}}$ of the previous linear section $n-1$. Equation (3.5) ensures that the output O_s lies on exactly one linear section n when component s is active.

We assume that an MILP model with fewer binary variables can be more efficiently solved. This assumption is often made in practice (Katz et al., 2020). The number of binary variables in the MILP model rises with the number of piecewise-affine sections. Thus, the objective of the data-driven model generation is to identify the minimal number of piecewise-affine sections L^{system} for a multi-energy system model with a given accuracy.

The resulting structure of the data-driven model-generation problem for multi-energy systems is the following:

min number of piecewise-affine sections in multi-energy system model (Eq. (3.6))
s.t. the multi-energy system model fulfills a given accuracy (Eq. (3.7)),
the multi-energy system model is fitted to measured data (Eq. (3.8)),
the component models are piecewise linear (Eq. (3.9)-(3.10)),
the piecewise-affine models are continuous (Eq. (3.11)-(3.14)),
equations to count all piecewise-affine sections (Eq. (3.15)-(3.16)),
equations to set variable bounds (Eq. (3.17)-(3.20)).

The mathematical formulation of the data-driven model-generation problem for multi-energy systems is given in Equations (3.6)-(3.20):

$$\min L^{\text{system}} = \sum_{s \in S} L_s \tag{3.6}$$

$$\text{s.t. } \Delta C^{\text{system}} \leq \delta^{\text{rel}} \cdot \sum_{s \in S} \sum_{d \in D} c_s^{\text{input}} \cdot I_{s,d}^{\text{data}}, \tag{3.7}$$

$$\Delta C^{\text{system}} = \sum_{s \in S} c_s^{\text{input}} \cdot \sqrt{\sum_{d \in D} (I_{s,d}^{\text{data}} - I_{s,d}^{\text{model}})^2}, \tag{3.8}$$

$$I_{s,d}^{\text{model}} = \sum_{n \in N_s^{\max}} \gamma_{s,n,d} \cdot (A_{s,n} \cdot O_{s,d}^{\text{data}} + B_{s,n}), \quad \forall\, s \in S, d \in D, \tag{3.9}$$

$$\sum_{n \in N_s^{\max}} \gamma_{s,n,d} = 1, \quad \forall\, s \in S, d \in D, \tag{3.10}$$

$$0 = \kappa_{s,n+1} \cdot \left[(A_{s,n+1} - A_{s,n}) \cdot O_{s,n}^{\text{ub}} + B_{s,n+1} - B_{s,n} \right], \quad \forall\, s \in S, n \in N_s^{\max}, \tag{3.11}$$

$$O_{s,n}^{\text{ub}} \geq O_{s,d}^{\text{data}} \cdot \gamma_{s,n,d}, \quad \forall\, s \in S, n \in N_s^{\max}, d \in D, \tag{3.12}$$

$$O_{s,n}^{\text{ub}} \cdot \gamma_{s,n+1,d} \leq O_{s,d}^{\text{data}} \cdot \gamma_{s,n+1,d}, \quad \forall\, s \in S, n \in N_s^{\max}, d \in D, \tag{3.13}$$

$$0 \leq (O_{s,n+1}^{\text{ub}} - O_{s,n}^{\text{ub}}) \cdot \kappa_{s,n+1}, \quad \forall\, s \in S, n \in N_s^{\max}, \tag{3.14}$$

$$\kappa_{s,n} \geq \frac{1}{|D|} \cdot \sum_{d \in D} \gamma_{s,n,d}, \quad \forall\, s \in S, n \in N_s^{\max}, \tag{3.15}$$

$$L_s = \sum_{n \in N_s^{\max}} \kappa_{s,n}, \quad \forall\, s \in S, \tag{3.16}$$

$$I_{s,d}^{\text{model}} \geq 0, \quad \forall\, s \in S, d \in D, \tag{3.17}$$

$$O_{s,n}^{\text{ub}} \geq 0, \quad \forall\, s \in S, n \in N_s^{\max}, \tag{3.18}$$

$$A_{s,n}, B_{s,n} \leq m \cdot \kappa_{s,n}, \quad \forall\, s \in S, n \in N_s^{\max}, \tag{3.19}$$

$$A_{s,n}, B_{s,n} \geq - m \cdot \kappa_{s,n}, \quad \forall\, s \in S, n \in N_s^{\max}. \tag{3.20}$$

In the following, we refer to the data-driven model generation problem for multi-energy systems (Equations (3.6)-(3.20)) as the actual problem. We state the variables and parameters of the actual problem in Table 3.1.

Table 3.1: List of all variables and parameters of the actual problem given in Equations (3.6)-(3.20).

Variables	$L^{\text{system}}, L_s, \Delta C^{\text{system}}, I_{s,d}^{\text{model}}, \gamma_{s,n,d}, A_{s,n}, B_{s,n}, O_{s,n}^{\text{ub}}, \kappa_{s,n}$
Parameters	$\delta^{\text{rel}}, c_s^{\text{input}}, I_{s,d}^{\text{data}}, O_{s,d}^{\text{data}}, \lvert D \rvert, m$

$d \in D$ is a measured data point. c_s^{input} is a cost-based weighting factor and depends on the type of input for component s. Different components of the multi-energy system may have different forms of input (e.g., gas for a boiler, but electricity for a compression chiller). By using cost-based weighting factors c_s^{input}, we convert the different forms of input to operating cost. We show in Section 3.2.1 how we determine cost-based weighting factors c_s^{input} for a typical multi-energy system. $N_s^{\max}$ is the set containing the maximum number of piecewise-affine sections allowed to model component s. The binary variable $\kappa_{s,n}$ denotes whether linear section n is used to model component s. The binary variable $\gamma_{s,n,d}$ denotes whether data point d is assigned to linear section n of component s. $A_{s,n}$ is the slope and $B_{s,n}$ is the intercept of linear section n in the piecewise-affine model of component s.

The objective of the actual problem is to minimize the number of piecewise-affine sections L^{system} within the multi-energy system model (Equation (3.6)). Constraints (3.7)-(3.8) restrict the sum of squared residuals of all data points d and components s to be smaller than the product of the predefined relative error of the multi-energy system δ^{rel} and the sum of the cost of all measured input data. Here, the sum of squared residuals of all data points d of component s are weighted by the cost-based weighting factor c_s. Constraints (3.9) evaluate the piecewise-affine models at each data point d for each component s. Constraints (3.10) ensure that each data point d

is assigned to exactly one linear section n for each component s. Constraints (3.11) force the piecewise-affine models to be continuous at the breakpoints. Constraints (3.12)-(3.14) define the variables for the upper bound $O_{s,n}^{\mathrm{ub}}$ of each linear section n and arrange the linear sections in ascending order. Constraints (3.15) ensure that the linear section n is chosen to model component s if at least one data point d is assigned to the linear section n. Constraints (3.16) sum up the number of chosen linear sections for each component s. Constraints (3.17)-(3.18) ensure $I_{s,d}^{\mathrm{model}}$ and $O_{s,n}^{\mathrm{ub}}$ to be positive variables. Constraints (3.19)-(3.20) assign the value 0 to the variables $A_{s,n}$ and $B_{s,n}$ if linear section n is not selected for component s. The constraints use a Big-M formulation with the Big-M value m.

The actual problem is an MINLP problem. The nonlinear character of the actual problem results from Equations (3.8), (3.9), (3.11), and (3.14). Solving the actual problem is computationally demanding. We implemented the actual problem in GAMS (GAMS Development Corporation, 2016) and tried to solve the actual problem with state-of-the-art MINLP solvers (SCIP (Gleixner et al., 2018), BARON (Tawarmalani and Sahinidis, 2005), DICOPT (Kocis and Grossmann, 1989) and BONMINH (Lougee-Heimer, 2003)). The MINLP solvers could not even find a feasible solution for a typically sized industrial multi-energy system (Section 3.3.5). The MINLP solvers ran without a time limit. Two solvers wrongly considered the problem infeasible, one solver terminated without a solution, and one reached an iteration limit. Thus, solving the actual problem is impractical in applications.

However, the actual problem can be rendered computationally feasible in two ways: One way is to linearize the actual problem (MINLP) to an MILP, based on the formulation by Yang et al. (2016). This linearization introduces a few shortcomings: Squared residuals can no longer be employed in an MILP (Equation (3.8)). Absolute residuals are calculated instead. Furthermore, the resulting piecewise-affine models are not continuous because the nonlinear continuity constraint cannot be considered in an MILP (Equation (3.11)). Thus, in general, the solution of the linearized problem is not a feasible solution to the actual problem. Recently, Kong and Maravelias (2020) and Rebennack and Krasko (2020) reformulated the nonlinear continuity constraint into a set of linear constraints. However, we show in Section 3.3.5 that the performance of the linearized problem is not always satisfying for practical applications, even if the continuity constraint is ignored. Therefore, there is a need for a solution method that provides a solution to the actual problem in a short time.

This need leads to the second way: decomposing the actual problem and solving the decomposed problem. For this purpose, we propose the decomposition method

AutoMoG in this work. AutoMoG provides a solution to the actual problem in a short time and, thus, is suitable for practical applications.

3.2 General workflow of AutoMoG

AutoMoG decomposes the actual problem (Equations (3.6)-(3.20)) to piecewise-linear regression problems for each component in a multi-energy system. AutoMoG iteratively increases the accuracy of the multi-energy system model by increasing the number of linear sections L^{system} (Figure 3.1).

In the following, we briefly outline the AutoMoG method before we explain the steps of AutoMoG in detail.

Step 1: For each component s, AutoMoG determines the operating range. The operating range of a component describes the feasible region of its model in a subsequent optimization. To determine the operating range of a component, AutoMoG calculates the convex hull (Barber et al., 1996) around the input data of the component. For this reason, the components' input data should ideally reflect its entire operating range. The edges of all convex hulls are added to the subsequent optimization problem as inequality constraints that limit the feasible region of the components' models.

Step 2: AutoMoG initializes the multi-energy system model by performing a linear regression with one section (one segment) for each component s, minimizing the mean squared error.

Step 3: AutoMoG checks the accuracy of the overall multi-energy system model (Section 3.2.1). For this purpose, the errors between model and measured data of all components are aggregated to the relative error of the overall multi-energy system $\Delta C^{\text{rel,System}}$. AutoMoG uses a cost-based weighting factor c_s^{input} to determine the impact of a component's error on the error of the multi-energy system model. The cost-based weighting factor c_s^{input} depends on a component's input because AutoMoG models the input as a function of the output (Equation (3.1)). We describe how to determine the cost-based weighting factors c_s^{input} in Section 3.2.1. AutoMoG terminates if the relative error of the multi-energy system model $\Delta C^{\text{rel,System}}$ is smaller than or equal to the allowed relative error δ^{rel}.

Instead of this accuracy check, it is also possible to limit the model complexity by specifying the number of piecewise-affine sections in the multi-energy system model. In this case, AutoMoG terminates if the pre-specified number of piecewise-affine sections is reached.

Step 4: If the relative error of the multi-energy system model $\Delta C^{\text{rel,System}}$ exceeds the allowed relative error δ^{rel}, AutoMoG calculates the next possible refinement for the component that was refined in the previous iteration. For this purpose, AutoMoG increases the number of piecewise-affine sections of the component by 1 and performs a least-squares regression between measured input and output data (Section 3.2.2). In the first iteration, AutoMoG calculates the next refinement for all components in the multi-energy system. As a result, the next possible refinement of each component is always known.

Step 5: AutoMoG checks the Corrected Akaike Information Criterion AIC_C (Hurvich and Tsai, 1993) to avoid overfitting. The Corrected Akaike Information Criterion AIC_C is used for model selection by capturing the trade-off between model accuracy and model complexity. If the information criterion AIC_C worsens for the refinement of a component, overfitting might occur. Thus, AutoMoG does not refine any component for which the information criterion AIC_C worsens. AutoMoG terminates if the information criterion AIC_C worsens for the refinement of all possible components. If AutoMoG terminates in Step 5, the allowed relative error δ^{rel} is not reached, but the measured data do not allow a more accurate model of the multi-energy system without the risk of overfitting.

Step 6: AutoMoG chooses one component to be refined based on the expected improvement in the overall error of the multi-energy system model (Section 3.2.4). Based on the calculated next refinements of all components, AutoMoG checks the improvement in the overall error of the multi-energy system model. The component with the greatest improvement in the overall error is chosen to be refined.

By the iteration of steps 3 to 6, AutoMoG increases the accuracy of the overall multi-energy system model until either the given accuracy is reached or all components in the multi-energy system would be overfitted when further refined. AutoMoG provides a fully parameterized MILP model of the multi-energy system. AutoMoG aims to find the minimal number of piecewise-affine sections L^{system} to represent the multi-energy system accurately. However, AutoMoG cannot guarantee to provide the model with the minimal number of piecewise-affine sections L^{system}.

In the following, we explain steps 3 to 6 of AutoMoG in detail.

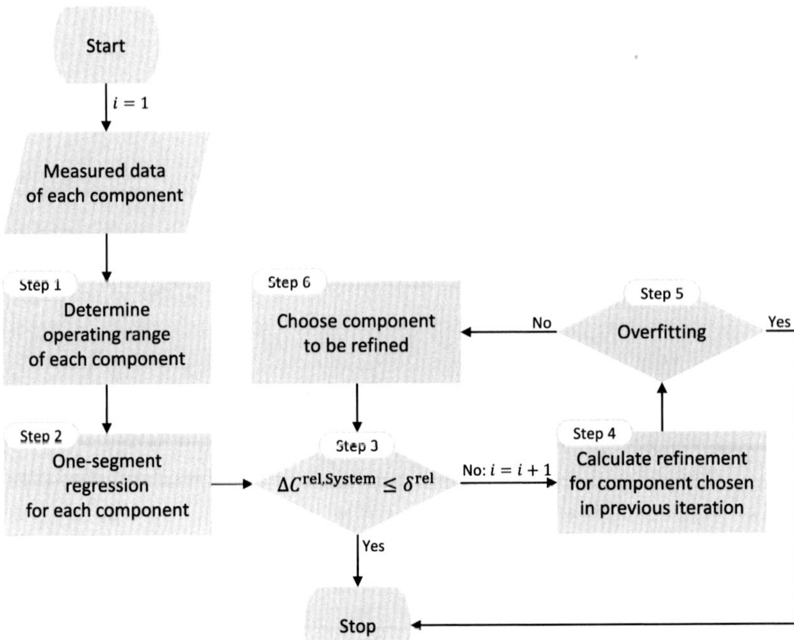

Figure 3.1: Proposed method AutoMoG for automated model generation using measured data. $\Delta C^{\text{rel,System}}$ is the relative error of the multi-energy system model (Section 3.2.1). δ^{rel} is the allowed relative error of the multi-energy system model.

3.2.1 Step 3: Accuracy measure on system level

After AutoMoG generates a model of each component in step 1, all component models are merged to a model of the multi-energy system. The accuracy of the multi-energy system model is then assessed. For a sufficiently accurate multi-energy system model, the relative error $\Delta C^{\text{rel,System}}$ shall be smaller than the allowed relative error δ^{rel} of the multi-energy system model:

$$\Delta C^{\text{rel,System}} \leq \delta^{\text{rel}} \quad . \tag{3.21}$$

The allowed relative error δ^{rel} is a user-specified parameter. The allowed relative error can be set to $\delta^{\text{rel}} = 0$ if the user is not able to specify an appropriate value. In this case,

AutoMoG provides an MILP model of the multi-energy system that is as accurate as possible without overfitting the input-output relationships of the components due to the use of the Corrected Akaike Information Criterion AIC_C (cf. Section 3.2.3).

The only information on the actual multi-energy system is the measured input and output data of each component. Thus, AutoMoG uses this information to calculate the relative error of the multi-energy system model $\Delta C^{\text{rel,System}}$ as

$$\Delta C^{\text{rel,System}} = \frac{\Delta C^{\text{system}}}{\sum_{s \in S} \sum_{d \in D} c_s^{\text{input}} \cdot I_{s,d}^{\text{data}}} \quad . \tag{3.22}$$

ΔC^{system} is the error of the multi-energy system model. $I_{s,d}^{\text{data}}$ is the measured input data of component s, and c_s^{input} is the cost based weighting factor of component s. The error of the multi-energy system model ΔC^{system} is the sum of the component model errors ΔC_s:

$$\Delta C^{\text{system}} = \sum_s \Delta C_s, \tag{3.23}$$

$$\text{with} \quad \Delta C_s = c_s^{\text{input}} \cdot \sqrt{\varepsilon_s}, \qquad \forall\, s \in S. \tag{3.24}$$

In the following, we explain why we use the sum of squared residuals ε_s and the cost-based weighting factors c_s^{input} to calculate the component model error ΔC_s (Equation (3.24)).

3.2.1.1 Sum of squared residuals ε_s

AutoMoG uses the sum of squared residuals ε_s to calculate the component model error ΔC_s of component s (Equation (3.24)). As a result, components with many data points tend to have a higher component model error ΔC_s and, thus, have a higher impact on the error of the multi-energy system model ΔC^{system} (Equation (3.23)). Thereby, AutoMoG takes into account that frequently used components are more important for the operation of the actual multi-energy system than rarely used components. However, components with many data points are not inherently more important. Thus, using the sum of squared residuals is only meaningful if the number of data points reflects the importance of the component compared to other components and not, e.g., only a lack of measured data. Preferentially, the data of all components is measured at the same time interval, using the same time step for the measurements. If the number of data points is known not to reflect the importance of a component, alternative error measures could be used, e.g., the mean squared error where the sum of squared residuals is divided by the number of data points for each component.

3.2.1.2 Cost-based weighting factors c_s^{input}

The obtained multi-energy system model is assumed to be used for economic optimization. To obtain a targeted model for economic optimization, AutoMoG assesses the component model errors ΔC_s in terms of operating cost.

However, measured data is commonly not available as operating cost. Instead, the amount of input and output energy is measured. Thus, AutoMoG converts the amount of input and output energy to operating cost using cost-based weighting factors c_s^{input} (Equation (3.24)). The cost-based weighting factors c_s^{input} enable balancing the component model errors ΔC_s in terms of cost. Thereby, AutoMoG incorporates the purpose of the model into the modeling process.

However, AutoMoG is not limited to generating models for economic optimization. Other weighting factors (e.g., primary energy factors or CO_2-eq.) can be implemented easily to obtain an optimization model targeted for other objective functions.

3.2.1.3 Determination of cost-based weighting factors c_s^{input} for a multi-energy system

The user has to provide a cost-based weighting factor for each energy form that is an input of at least one component in the multi-energy system. In the following, we illustrate the determination of cost-based weighting factors for a multi-energy system with gas-driven boilers and CHP engines, heat-driven absorption chillers, and electricity-driven compression chillers. Thus, cost-based weighting factors are required for gas (input for boilers and CHP engines), heat (input for absorption chillers), and electricity (input for compression chillers).

For components driven by energy forms purchased from an external grid (e.g., gas and electricity), we propose choosing the specific prices of the energy forms (c^{gas} and c^{el}) as cost-based weighting factors.

For components driven by energy forms that are not purchased from an external grid (e.g., heat), we need to determine a cost-based weighting factor that approximates the specific cost for this energy form in the multi-energy system. In the given example, heat is supplied by different components in the multi-energy system, e.g., by CHP engines or boilers. We want to calculate one cost-based weighting factor for heat. For this purpose, we average the cost of all heat-supplying components. This procedure needs to be applied for every energy form that cannot be purchased directly but is used within the energy system. The procedure can also be adapted when using AutoMoG to generate models of other systems, for example, for intermediate chemicals that are transformed into desired fuels in chemical plants.

For heat supplied by boiler b, the component-specific cost c_b^{heat} is taken from the operation of nominal load:

$$c_b^{\text{heat}} = \frac{c^{\text{gas}}}{\eta_b^{\text{nom}}}, \qquad \forall\, b \in B. \tag{3.25}$$

η_b^{nom} is the nominal efficiency of boiler b extracted from measured data. For this extraction, we search the data point with the maximum heat output. This maximum heat output is divided by the corresponding gas input to calculate the nominal efficiency η_b^{nom}.

For heat supplied by CHP engine chp, the component-specific cost c_{chp}^{heat} is calculated using the energetic method from The Association of German Engineers (2008). The energetic method allocates the cost of purchased gas to the supplied heat and the supplied electricity of the CHP engine based on the amount of supplied thermal and electrical energy:

$$c_{chp}^{\text{heat}} = \frac{c^{\text{gas}}}{\eta_{chp}^{\text{heat}} + \eta_{chp}^{\text{el}}}, \qquad \forall\, chp \in CHP. \tag{3.26}$$

The thermal efficiency η_{chp}^{heat} and the electrical efficiency η_{chp}^{el} of the CHP engine are extracted from measured data in the same manner as the nominal efficiency η_b^{nom} of boiler b.

The overall cost-based weighting factor c^{heat} for heat in the multi-energy system is calculated from the component-specific cost for heat supplied by each boiler and CHP engine:

$$c^{\text{heat}} = \sum_b c_b^{\text{heat}} \cdot \frac{Q_b}{Q^{\text{system}}} + \sum_{chp} c_{chp}^{\text{heat}} \cdot \frac{Q_{chp}}{Q^{\text{system}}}, \tag{3.27}$$

with Q_b and Q_{chp} being the heat amount supplied by boiler b and CHP engine chp, respectively. Q^{system} is the overall heat amount supplied in the multi-energy system. Q_b, Q_{chp}, and Q^{system} are extracted from measured data. To calculate the cost-based weighting factor c^{heat} for heat, the component-specific cost for heat is weighted by the amount of supplied heat from the corresponding component.

With the determined cost-based weighting factors, we can calculate the relative error of the multi-energy system $\Delta C^{\text{rel,System}}$ in Equation (3.22). However, the proposed procedure to determine the cost-based weighting factors is not exact. Still, we find that the cost-based weighting factors are important to consider and therefore discuss their impact (Section 3.3.4).

With the cost-based weighting factors, AutoMoG is able to check the accuracy of the multi-energy system model using Equation (3.21). If the multi-energy system model does not fulfill the desired accuracy measure in Equation (3.21), AutoMoG increases the number of piecewise-affine sections L^{system} in the multi-energy system model. Thus, AutoMoG refines one component to decrease the relative error of the multi-energy system model $\Delta C^{\text{rel,System}}$.

3.2.2 Step 4: Piecewise-linear regression for each component

In the first iteration, AutoMoG solves the piecewise-affine regression problem for each component s with 1 linear section (Step 2). In each subsequent iteration i, AutoMoG chooses one component s to be refined by allowing one more linear section in the piecewise-linear regression problem of component s:

$$|N_{s,i+1}| = |N_{s,i}| + 1, \tag{3.28}$$

with $N_{s,i}$ being the set of piecewise-affine sections used to model component s in iteration i.

AutoMoG solves the piecewise-linear regression problem for each component $s \in S$ by minimizing the sum of squared residuals ε_s. The squared residuals between the modeled input $I_{s,d}^{\text{model}}$ and the measured input $I_{s,d}^{\text{data}}$ are summed up for all measured data points $d \in D$, finally resulting in the following optimization problem:

$$\min_{A_{s,n}, B_{s,n}, O_{s,n}^{\text{ub}}} \quad \varepsilon_s = \sum_{d \in D} \left(I_{s,d}^{\text{data}} - I_{s,d}^{\text{model}} \right)^2 \tag{3.29}$$

$$\text{s.t.} \quad I_{s,d}^{\text{model}} = \sum_{n \in N_{s,i}} \gamma_{s,n,d} \cdot (A_{s,n} \cdot O_{s,d}^{\text{data}} + B_{s,n}), \quad \forall d \in D, \tag{3.9}$$

$$\sum_{n \in N_{s,i}} \gamma_{s,n,d} = 1, \quad \forall d \in D, \tag{3.10}$$

$$0 = (A_{s,n+1} - A_{s,n}) \cdot O_{s,n}^{\text{ub}} + B_{s,n+1} - B_{s,n}, \quad \forall n \in N_{s,i}, \tag{3.11}$$

$$O_{s,n}^{\text{ub}} \geq O_{s,d}^{\text{data}} \cdot \gamma_{s,n,d}, \quad \forall n \in N_{s,i}, d \in D, \tag{3.12}$$

$$O_{s,n}^{\text{ub}} \cdot \gamma_{s,n+1,d} \leq O_{s,d}^{\text{data}} \cdot \gamma_{s,n+1,d}, \quad \forall n \in N_{s,i}, d \in D, \tag{3.13}$$

$$I_{s,d}^{\text{model}} \geq 0, \quad \forall d \in D, \tag{3.17}$$

$$O_{s,n}^{\text{ub}} \geq 0, \quad \forall n \in N_{s,i}. \tag{3.18}$$

The constraints (3.9)-(3.13), (3.17), and (3.18) of the piecewise-linear regression problem are the same as in the actual problem. The objective function (3.29) of the piecewise-linear regression problem is the sum of squared residuals ε_s for each component s, derived from constraint (3.8) of the actual problem. From this regression problem, AutoMoG obtains the parameters of the piecewise-affine model $A_{s,n}$, $B_{s,n}$ and the positions of the breakpoints $O^{\mathrm{ub}}_{s,n}$ of each component s. The set of piecewise-affine sections $N_{s,i}$ is fixed for each piecewise-linear regression problem.

Methods for solving the piecewise-linear regression problem are available in the literature (Kong and Maravelias, 2020; Rebennack and Krasko, 2020; Yang et al., 2016; Zhang et al., 2016; Camponogara and Nazari, 2015). Here, the piecewise-linear regression problem is an MINLP that is solved by applying an existing Matlab-Toolbox (D'Errico, 2009). The Matlab-Toolbox reformulates the MINLP into NLP subproblems and solves the subproblems with a local NLP solver. The Matlab-Toolbox initializes the positions of the breakpoints for the piecewise-affine functions equidistantly and calculates the sum of squared residuals ε_s. After the initialization, the Matlab-Toolbox minimizes the sum of squared residuals ε_s by iteratively changing the positions of the breakpoints. The piecewise-affine model of each component s is constrained to be continuous at the breakpoints (Equation (3.11)). However, AutoMoG cannot guarantee to find the globally optimal positions of the breakpoints as it uses a local NLP solver for the NLP subproblems. For proof of optimality, the approaches of Kong and Maravelias (2020) and Rebennack and Krasko (2020) can be used instead of the Matlab-Toolbox.

Thus, step 4 of AutoMoG provides a continuous piecewise-affine model of each component in the multi-energy system.

3.2.3 Step 5: Avoid overfitting with the Corrected Akaike Information Criterion $\mathrm{AIC_C}$

In Step 5, AutoMoG aims to avoid overfitting. For this purpose, AutoMoG checks the Corrected Akaike Information Criterion $\mathrm{AIC_C}$ for each component before choosing a component to be refined.

Information criteria have been developed to select the most suitable model for a data set. The model with the lowest value of the used information criterion is selected. Widely known information criteria are, e.g., the Akaike Information Criterion AIC (Akaike, 1974) or the Bayesian Information Criterion BIC (Stoica and Selén, 2004). In AutoMoG, we use the Corrected Akaike Information Criterion $\mathrm{AIC_C}$ (Hurvich and

Tsai, 1993) since it is an extension of the AIC suitable for small sample sizes. However, other information criteria can be implemented easily in AutoMoG.

The Corrected Akaike Information Criterion AIC_C has been proposed by Burnham and Anderson (2003) with the assumption of normally distributed errors in the measured data as follows:

$$\text{AIC}_{\text{C},s,i} = d \cdot \ln\left(\frac{\varepsilon_s}{d}\right) + 2K_{s,i} + \frac{2K_{s,i} \cdot (K_{s,i}+1)}{d - K_{s,i+1} - 1}, \quad \forall s \in S, \tag{3.30}$$

with d being the number of data points and ε_s being the sum of squared residuals. $K_{s,i}$ is the total number of regression parameters used to describe the model. The first term of the sum in Equation (3.30) rewards model accuracy, whereas the other terms of the sum penalize model complexity.

AutoMoG uses the Corrected Akaike Information Criterion $\text{AIC}_{\text{C},s,i}$ to compare the model of component s from iteration i to its refined model from iteration $i+1$. Thus, if for component s

$$\text{AIC}_{\text{C},s,i+1} \geq \text{AIC}_{\text{C},s,i} \tag{3.31}$$

holds, the improvement in model accuracy does not overcome the increase in model complexity from iteration i to iteration $i+1$. Thus, further model refinement of component s would probably risk overfitting. Consequently, AutoMoG does not refine any components for which the Corrected Akaike Information Criterion $\text{AIC}_{\text{C},s,i}$ increases. Instead, AutoMoG refines only one of the components for which the Corrected Akaike Information Criterion $\text{AIC}_{\text{C},s,i}$ decreases.

AutoMoG terminates if the Corrected Akaike Information Criterion $\text{AIC}_{\text{C},s,i}$ increases for every component, even if the allowed relative error δ^{rel} is not reached (Figure 3.1). If there is at least one component for which the Corrected Akaike Information Criterion $\text{AIC}_{\text{C},s,i}$ decreases, AutoMoG proceeds with step 6.

3.2.4 Step 6: Choose component to be refined

In step 6, the component to be refined is chosen by identifying the maximum error reduction. The maximum error reduction is the maximum difference between the component model error $\Delta C_s(N_{s,i})$ and the component model error in the next refinement $\Delta C_s(N_{s,i+1})$:

$$\max_s \left(\Delta C_s(N_{s,i}) - \Delta C_s(N_{s,i+1})\right). \tag{3.32}$$

Thus, the component model error of the refinement $\Delta C_s(N_{s,i+1})$ has to be known already. Consequently, in the first iteration, AutoMoG has to solve the piecewise-linear regression problem with two linear sections for each component in step 4. In all subsequent iterations, only one additional piecewise-linear regression problem has to be solved for the component s refined in the previous iteration.

After choosing the component to be refined in step 6, AutoMoG proceeds with step 3 (Figure 3.1). AutoMoG terminates once either the allowed relative error δ^{rel} is reached, or all components in the multi-energy system would be overfitted when further refined.

3.3 Case Study: Multi-energy system from literature

In this section, we apply AutoMoG to a case study based on Goderbauer et al. (2016). Section 3.3.1 describes the case study. Section 3.3.2 presents the results of AutoMoG. As a benchmark approach, the common approach is employed where each component is modeled independently. In Section 3.3.3, we test the performance of the generated multi-energy system model in operational optimization. Section 3.3.4 shows a sensitivity analysis for the cost-based weighting factors used in the case study. Section 3.3.5 compares the performance of the actual problem and the linearized problem to AutoMoG.

3.3.1 Description of the multi-energy system

The case study is based on Goderbauer et al. (2016), who studied a real-world multi-energy system (Figure 3.2). The multi-energy system consists of two identical boilers, a small and a large CHP engine, a small and a large absorption chiller, and two identical compression chillers. In their model, Goderbauer et al. (2016) use nonlinear input-output relationships for all components. The boilers, absorption chillers, and compression chillers are modeled with one input and one output each: The boilers are driven by gas and supply heat; the absorption chillers are driven by heat and supply cooling; the compression chillers are driven by electricity and supply cooling. Thus, for each of these components, one input-output relationship is required.

However, the CHP engines have one input (gas) and two outputs (heat and electricity). Thus, two input-output relationships are required to model a CHP engine. Goderbauer et al. (2016) model both gas input and electricity output as a function of the heat output:

$$I_{chp,d}^{\text{model}} = f\left(O_{chp,d}^{\text{heat}}\right), \qquad \forall\ chp \in CHP,\ d \in D, \tag{3.33}$$

$$O_{chp,d}^{\text{el,model}} = f\left(O_{chp,d}^{\text{heat}}\right), \qquad \forall\ chp \in CHP,\ d \in D. \tag{3.34}$$

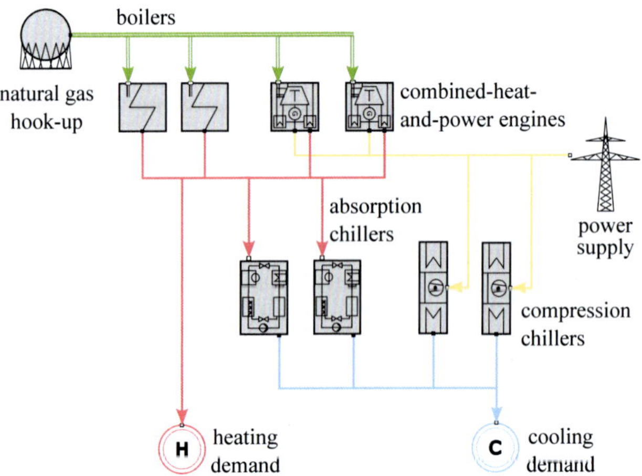

Figure 3.2: Flowsheet of the multi-energy system in the case study. This multi-energy system is modeled by the proposed method AutoMoG.

Following this modeling approach for CHP engines, AutoMoG uses as cost-based weighting factors the specific gas price c^{gas} for the gas input $I_{chp,d}^{\text{model}}$ and the specific electricity price c^{el} for the electricity output $O_{chp,d}^{\text{el,model}}$.

In total, ten piecewise-affine models are required to describe the input-output relationships of the eight components in the case study (one for each boiler, absorption chiller and compression chiller, two for each CHP engine).

To generate a multi-energy system model, AutoMoG requires measured input and output data of all modeled components. To obtain the required input and output data in the case study, we simulate 100 load cases of the multi-energy system with the nonlinear input-output relationships from Goderbauer et al. (2016). The 100 load cases are created by aggregating the demand time-series of heat, cooling, and electricity from Goderbauer et al. (2016). The demand time series of one year with a resolution of 1 h is aggregated to 100 typical time steps, using k-medoids (Kaufman and Rousseeuw, 1987). The simulation of the 100 typical time steps provides the required input and output data of each component for 100 load cases. We use the input and output data of each component from the simulation and add normally distributed noise in a range of ±5 % of the simulated values. The thus obtained noisy data are used as measured input and output data in the case study.

To apply AutoMoG, we have to choose a value for the allowed relative error δ^{rel} of the multi-energy system model. For this purpose, we calculate the relative error of the multi-energy system model $\Delta C^{\text{rel,System}}$ with the nonlinear input-output relationships from Goderbauer et al. (2016) compared to the obtained noisy data. We choose this relative error of the multi-energy system model as the allowed relative error δ^{rel}. Aiming for a higher accuracy than the actual functional relationship is not reasonable. However, if a meaningful allowed relative error δ^{rel} is not available, we recommend setting the allowed relative error $\delta^{\text{rel}} = 0$. In such cases, AutoMoG still provides a meaningful model as the Corrected Akaike Information Criterion AIC_C discourages overfitting (Section 3.2.3).

3.3.2 Modeling the multi-energy system using AutoMoG

The AutoMoG method is applied to the case study (Section 3.3.1). AutoMoG provides a multi-energy system model for the case study that reaches the allowed relative error δ^{rel}. To reach the allowed relative error δ^{rel}, AutoMoG needs four iterations and less than one minute. The multi-energy system model uses 14 piecewise-affine sections to describe the input-output relationships of all components. The solution provided by AutoMoG is a feasible solution for the actual problem (Equations (3.6)-(3.20)).

In common practice, each component of a multi-energy system is modeled independently. Thus, we compare the results of AutoMoG to the independent modeling of each component. In independent modeling of each component, the relative error of the multi-energy system $\Delta C^{\text{rel,System}}$ cannot be evaluated during model generation. Thus, to ensure that the relative error of the multi-energy system $\Delta C^{\text{rel,System}}$ is lower than the allowed relative error δ^{rel}, the relative error of each component ΔI_s^{rel} has to be lower or equal than the allowed relative error δ^{rel}:

$$\Delta I_s^{\text{rel}} = \frac{\sqrt{\varepsilon_s}}{\sum_d I_{s,d}^{\text{data}}} \leq \delta^{\text{rel}}, \quad \forall s \in S, \tag{3.35}$$

with ε_s being the sum of squared residuals of component s and $I_{s,d}^{\text{data}}$ being the input of component s in data point d. In independent modeling, the only way to guarantee the fulfillment of Equation (3.35) is to ignore the Corrected Akaike Information Criterion AIC_C. However, ignoring the Corrected Akaike Information Criterion AIC_C may lead to overfitting. Thus, we generate two models of the multi-energy system with independent modeling:

- Independent modeling: we ignore the Corrected Akaike Information Criterion AIC_C
- Independent modeling (AIC_C): we apply the Corrected Akaike Information Criterion AIC_C.

Note that the cost-based weighting factors c_s are not used for independent modeling because the relative errors of the components $\Delta I_s^{\mathrm{rel}}$ are not compared to each other.

In the case study, independent modeling needs 31 piecewise-affine sections to reach $\Delta I_s^{rel} \leq \delta^{\mathrm{rel}}$ for each component s. Independent modeling ($\mathrm{AIC_C}$) uses 23 piecewise-affine sections and does not reach $\Delta I_s^{rel} \leq \delta^{\mathrm{rel}}$ for two components. Thus, AutoMoG provides an MILP model of the multi-energy system that needs significantly fewer piecewise-affine sections than both models from independent modeling (14 vs. 31 and 23, respectively). Still, the multi-energy system model provided by AutoMoG is sufficiently accurate and satisfies the error criterion for the multi-energy system $\Delta C^{\mathrm{rel,System}} \leq \delta^{\mathrm{rel}}$.

Figure 3.3 exemplary shows the models of AutoMoG and independent modeling for a compression chiller and the large CHP engine. The models are compared to the original models from Goderbauer et al. (2016) and the measured data. AutoMoG uses two linear sections to represent the input-output relationship of the compression chiller (cf. Figure 3.3 (a)), whereas independent modeling uses three linear sections in both models (cf. Figure 3.3 (c) and (e)). The three models seem appropriate to represent the input-output relationship of the compression chiller, and from visual inspect, none of the models is obviously the better model. However, the accuracy measured on the system level in AutoMoG reveals that two linear sections are sufficient for the model of the compression chiller.

For the model of the large CHP engine, AutoMoG uses one linear section, whereas independent modeling uses ten linear sections and, thereby, obviously overfits the input-output relationship of the large CHP engine (cf. Figure 3.3 (d)). The independent modeling leads to overfitting due to the need to reach the allowed relative error δ^{rel} for each component. If we discourage overfitting by the Corrected Akaike Information Criterion $\mathrm{AIC_C}$ in independent modeling, the model of the large CHP engine is modeled with three linear sections (cf. Figure 3.3 (f)). This finding shows that the Corrected Akaike Information Criterion $\mathrm{AIC_C}$ is powerful in preventing overfitting. Still, even with the Corrected Akaike Information Criterion $\mathrm{AIC_C}$ independent modeling increases model complexity strongly compared to AutoMoG (one vs. three linear sections).

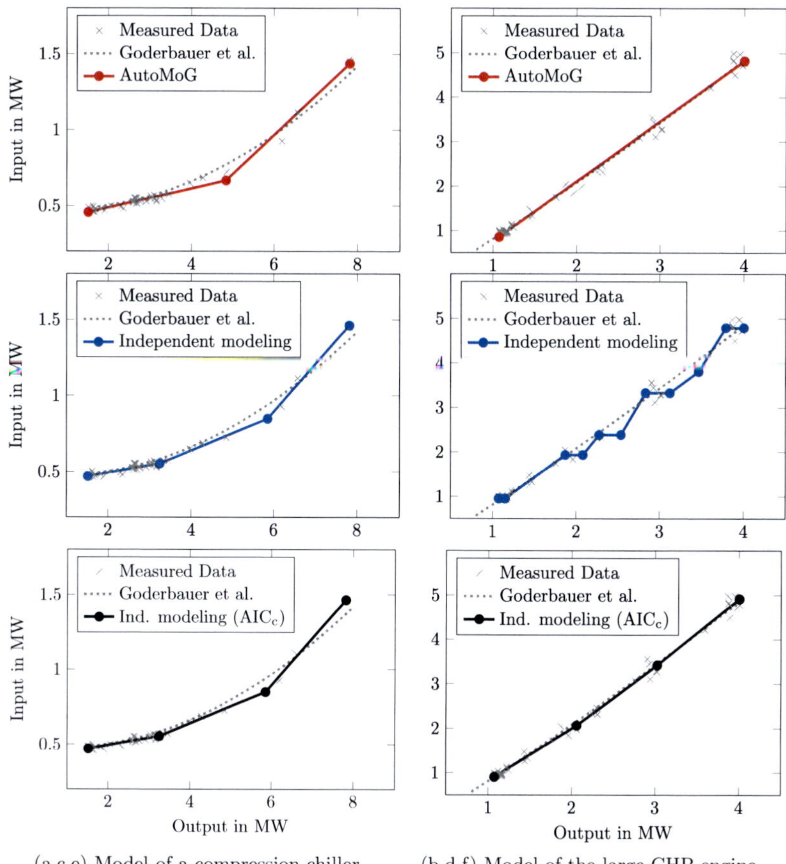

(a,c,e) Model of a compression chiller.

(b,d,f) Model of the large CHP engine.

Figure 3.3: Derived models of a compression chiller (a,c,e) and the large CHP engine (b,d,f) compared to the original models from Goderbauer et al. (2016) (grey dashed line) and the measured data (grey crosses). The AutoMoG models are shown in red (a,b), and the models from independent modeling of each component are shown in blue (c,d), and black, respectively (e,f). Independent modeling of each component shows overfitting for the large CHP engine (d). The Corrected Akaike Information Criterion AIC_C can prevent overfitting in independent modeling. Still, AutoMoG uses fewer linear sections to model the large CHP engine.

3.3.3 Model performance in operational optimization

The multi-energy system model is generated for efficient operational optimization. Thus, we test the model performance by analyzing the computation times of the models.

For this purpose, we solve five instances of operational optimization problems with the models from AutoMoG, independent modeling, and independent modeling (AIC_C).

The original demand time-series of heat, cooling, and electricity from Goderbauer et al. (2016) consist of one year with an hourly resolution. Thus, the original demand time-series contains 8760 time steps. We create five instances of the original demand time series with latin-hypercube sampling (McKay et al., 2000) and variations of ±5 % of the original demands.

The AutoMoG model is the fastest model for all instances (cf. Table 3.2). As expected, the model from independent modeling is the slowest for all instances as it contains the most piecewise-affine sections and thus the most binary variables compared to the other two models. On average, the AutoMoG model solves the operational optimization problems more than 50 times faster than independent modeling. Even compared to independent modeling (AIC_C), the AutoMoG model solves more than 15 times faster on average. The computation times show that AutoMoG generates efficient models for operational optimization.

Table 3.2: Computation times of operational optimization for all five instances and all three models. We implemented the operational optimization problems in Python and solved them with Gurobi 9.0.0 (Gurobi Optimization, 2020).

Instance	1	2	3	4	5	∅
AutoMoG	32 s	31 s	32 s	31 s	31 s	31 s
Independent modeling	1595 s	1674 s	1566 s	1546 s	1505 s	1577 s
Ind. modeling (AIC_C)	438 s	533 s	603 s	452 s	588 s	523 s

Additionally, we test the model performance by analyzing how well the models predict the operating cost $OPEX$. For this purpose, we perform a 5-fold cross-validation for the already used 100 operating points.

For the 5-fold cross-validation, the 100 operating points are split into five data sets of equal size. Afterwards, four data sets are used for model generation, and the remaining data set is used as the test set. Thus, each data set is used once as the test set.

As accuracy measure for the cross-validation, we use the relative difference in operating cost for each operating point d $\Delta OPEX_d^{\text{rel}}$:

$$\Delta OPEX_d^{\text{rel}} = \left| \frac{OPEX_d^{\text{data}} - OPEX_d^{\text{model}}}{OPEX_d^{\text{data}}} \right|, \quad \forall d \in D. \tag{3.36}$$

with $OPEX_d^{\text{data}}$ being the operating cost of the multi-energy system according to the measured data and $OPEX_d^{\text{model}}$ being the operating cost of the multi-energy system according to the tested model.

The 5-fold cross-validation is applied to the AutoMoG model and both models from independent modeling of each component (cf. Table 3.3).

Table 3.3: Results of the 5-fold cross-validation for the case study. $\overline{\Delta OPEX^{\text{rel}}}$ is the mean relative difference in operating cost between model and data. N is the number of piecewise-affine sections in the multi-energy system model.

Test set		1	2	3	4	5	∅
	AutoMoG	0.93	0.91	0.76	0.55	0.74	0.78
$\overline{\Delta OPEX^{\text{rel}}}/\%$	Independent modeling	0.95	0.95	1.11	0.71	1.19	0.98
	Ind. modeling (AIC$_\text{C}$)	0.79	0.78	0.86	0.77	0.70	0.78
	AutoMoG	16	15	17	16	16	16
N	Independent modeling	34	35	40	32	45	37.2
	Ind. modeling (AIC$_\text{C}$)	26	21	24	27	25	24.6

The results of the 5-fold cross-validation show that AutoMoG uses fewer piecewise-affine sections in each test set (16 linear sections on average in the models of AutoMoG compared to 37.2 on average in independent modeling and 24.6 on average in independent modeling (AIC$_\text{C}$)). However, still, AutoMoG reduces the mean relative difference in operating cost between model and data $\overline{\Delta OPEX^{\text{rel}}}$ by approx. 20 % from 0.98 % to 0.78 % compared to independent modeling. Furthermore, AutoMoG reaches the same mean relative difference in operating cost as independent modeling (AIC$_\text{C}$), although AutoMoG uses fewer piecewise-affine sections. Thus, in the case study, it is not necessary to model each component as accurately as possible to reach an accurate representation of the multi-energy system.

3.3.4 Sensitivity analysis of the cost-based weighting factors

In this section, we perform a sensitivity analysis of the cost-based weighting factors to analyze their importance. At the same time, we find that the cost-based weighting factors do not need to be calculated with high accuracy.

The cost-based weighting factors essentially sort the components by importance. In the case study (Section 3.3.1 and Section 3.3.2), we use as weighting factors for gas $c^{\text{gas}} = 0.06\,€/\text{kWh}$, for electricity $c^{\text{el}} = 0.16\,€/\text{kWh}$ and for heat $c^{\text{heat}} = 0.071\,€/\text{kWh}$. Thus, the errors of components that use electricity as input are weighted more than two times higher than the errors of components that use gas or heat as an input. AutoMoG needs four iterations and, thus, performs four refinements in total for the ten piecewise-affine models. The four refinements are: 1. large absorption chiller, 2. small absorption chiller, 3. a compression chiller, 4. a boiler.

For a sensitivity analysis of the cost-based weighting factors, we change each of the three weighting factors by $\pm 20\,\%$ and run AutoMoG once for each changed weighting factor. In each of the six additional runs, AutoMoG refines the same four components. However, the order of refinement changes when we increase the weighting factor for electricity to $c^{\text{el}} = 0.192\,€/\text{kWh}$ or reduce the weighting factor for heat to $c^{\text{heat}} = 0.057\,€/\text{kWh}$. In these two cases, a compression chiller is refined before the small absorption chiller is refined.

In another additional run, we ignore the weighting factors. Thus, the errors of all components have the same weight, independently of the input of the components. In this case, AutoMoG refines the models of the following four components in the following order: 1. large absorption chiller, 2. small absorption chiller, 3. a boiler, 4. another boiler. We can see that the compression chiller is only refined when we use the cost-based weighting factors. Thus, the cost-based weighting factors show that economically it makes sense to refine the compression chiller since electricity is more valuable than gas. If we ignore the weighting factors, different components are refined. Refining different components leads to a different model of the overall multi-energy system.

The sensitivity analysis shows that the weighting factors decide which components are refined in which order. However, changing the weighting factors by $\pm 20\,\%$ does not fundamentally change the results of the case study, which leads us to the conclusion that a rough estimation of the weighting factors is sufficient. In summary, we conclude that cost-based weighting factors are important if their values significantly differ for the input energy forms.

3.3.5 Computational study of the actual problem

AutoMoG is essentially a primal heuristic for the actual problem (Equations (3.6)-(3.20)) using problem decomposition to find a good feasible solution. However, there is no proof that the AutoMoG solution is optimal. Furthermore, the AutoMoG solution is only feasible for the actual problem if AutoMoG terminates because the allowed relative error δ^{rel} is reached. The allowed relative error δ^{rel} is not reached if AutoMoG stops because all components would be overfitted when further refined. In this case, the AutoMoG solution is infeasible for the actual problem because the allowed relative error δ^{rel} cannot be reached, and the corresponding constraint (3.7)) is violated.

To assess the solution quality and the performance of the AutoMoG method, we implemented the actual problem (Equations (3.6)-(3.20)) and the linearization of the actual problem (cf. Section 3.1) for the case study in GAMS. In the linearization of the actual problem, we followed the approach of Yang et al. (2016), i.e., we used absolute errors instead of squared residuals and ignored the constraints that enforce continuity at the breakpoints of the piecewise-affine functions. Thus, the linearization does not guarantee continuity at the breakpoints, which leads to different component models. Still, we learn about the computational effort of the MILP approach compared to AutoMoG.

To solve the actual problem, we used the MINLP solvers SCIP (Gleixner et al., 2018), BARON (Tawarmalani and Sahinidis, 2005), DICOPT (Kocis and Grossmann, 1989), and BONMINH (Lougee-Heimer, 2003). All MINLP solvers were used with default subsolvers and default settings, if not stated otherwise. The default subsolvers and default settings can be found in the solver manuals of the GAMS documentation (GAMS Development Corporation, 2016). Each run of the actual problem with an MINLP solver was executed without a time limit.

None of the used MINLP solvers were able to find any feasible solution for the actual problem (cf. Table 3.4). The solvers SCIP and BARON even identified the actual problem as infeasible. However, we showed that the AutoMoG solution is feasible for the actual problem by using the AutoMoG solution as a starting solution for SCIP, BARON, and DICOPT. Even with the AutoMoG solution as a starting solution, none of the used MINLP solvers was able to find another solution than the AutoMoG solution (cf. Table 3.4).

Table 3.4: Results of solving the actual problem with different MINLP solvers without starting solution and with the AutoMoG solution as starting solution.

Solver	No starting solution	Starting solution from AutoMoG
SCIP	MINLP is infeasible	Starting solution is feasible Starting solution is optimal
BARON	MINLP is infeasible	Starting solution is feasible Starting solution is optimal
DICOPT	No feasible solution found within Iteration Limit of 200	Starting solution is feasible No better solution found within Iteration Limit of 200
BONMINH	Terminated by solver	-

Thus, solving the actual problem with standard MINLP solvers is impractical and even impossible for the present case study.

To solve the linearized problem, we used CPLEX 12.6.0.1 (IBM Corporation, 2017) with a time limit of 48 h. CPLEX reached the time limit with a feasible solution of 17 piecewise-affine sections for the multi-energy system model (the AutoMoG solution uses 14 piecewise-affine sections).

In Figure 3.4, we compare the models of the large CHP engine and an absorption chiller derived by AutoMoG and by the linearized problem. Within the time limit of 48 h, the best model found by the linearized problem has three linear sections for the large CHP engine and shows discontinuities (cf. Figure 3.4 (c)), whereas AutoMoG finds a model for the large CHP engine with one linear section within 1 min (cf. Figure 3.4 (a)). The AutoMoG model of the large CHP engine is obviously more appropriate to represent the input-output relationship of the large CHP engine than the model derived by the linearized problem. For the model of the absorption chiller, the solution of the linearized problem uses three linear sections (cf. Figure 3.4 (d)), whereas AutoMoG uses two linear sections (cf. Figure 3.4 (b)).

In summary, even with the stated simplifications, we could not find a better solution of the linearized problem within 48 h than the solution AutoMoG finds within 1 min. In contrast to the AutoMoG solution, the solution of the linearized problem is not feasible for the actual problem because the solution of the linearized problem shows discontinuities at the breakpoints of the piecewise-affine models (cf. Figure 3.4). If we used the approaches of Kong and Maravelias (2020) and Rebennack and Krasko (2020) for the linearized problem, we could overcome the problem of discontinuities at the breakpoints. Since these methods further constrain the solutions presented in this work, we do not expect a significantly improved performance for the linearized problem. However, it would certainly be interesting to explore these methods in future work. Neither the actual problem nor the linearized problem is applicable in practice. In summary, we showed that the method AutoMoG provides an accurate and computationally efficient model of the multi-energy system in a short time.

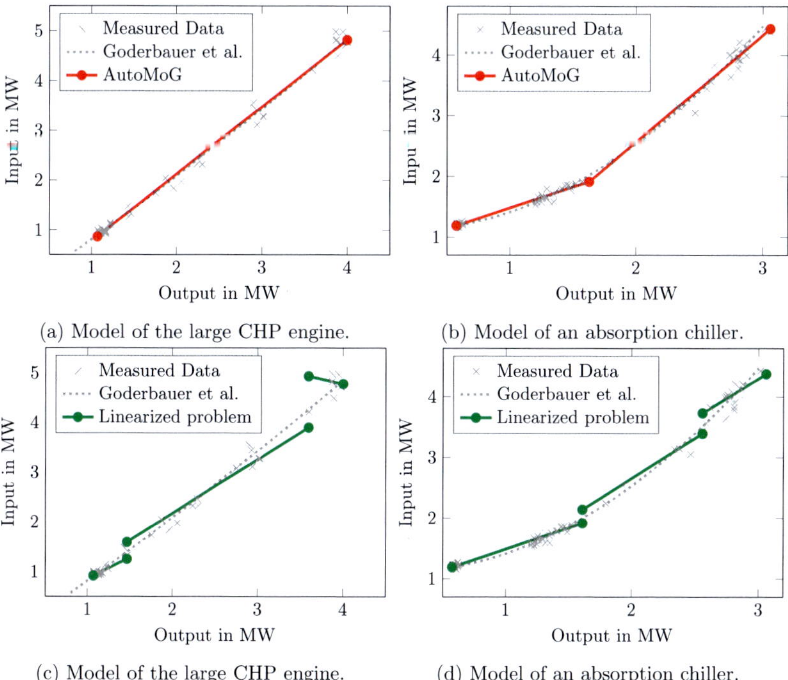

(a) Model of the large CHP engine.

(b) Model of an absorption chiller.

(c) Model of the large CHP engine.

(d) Model of an absorption chiller.

Figure 3.4: Derived model of the large CHP engine and an absorption chiller from AutoMoG (red) and from the linearized problem (green). Furthermore, the original models from Goderbauer et al. (2016) (grey dashed line) and the measured data (grey crosses) are presented. The solution of the linearized problem uses more linear sections and shows discontinuities at the breakpoints. Thus, the solution of the linearized problem is not a feasible solution to the actual problem.

3.4 Conclusion

The AutoMoG method is proposed for automated data-driven model generation of multi-energy systems using piecewise-linear regression. AutoMoG decomposes the data-driven model generation problem of a multi-energy system to one model generation problem of each component. Still, the error of model generation is evaluated for the overall multi-energy system. For this purpose, AutoMoG uses cost-based weighting factors to balance the errors caused by each component model. Through the decomposition, AutoMoG provides an accurate solution for the data-driven model generation problem of the multi-energy system in a short time. The solution provided by AutoMoG is an MILP model of the multi-energy system that is usable for optimization.

In the case study, AutoMoG needs significantly fewer piecewise-affine sections (57 % on average) to reach an allowed model error compared to the commonly employed independent modeling of each component. As a result, the model of AutoMoG solves the operational optimization more than 50 times faster on average. Still, the models from AutoMoG are more accurate (20 % on average) in terms of operating cost than the models provided by independent modeling of each component.

The results of the case study show that it is not mandatory to model each component with high accuracy to reach an accurate multi-energy system model. Instead, each component's modeling error should be seen in the context of the multi-energy system model.

The proposed method AutoMoG is only applicable if measured input and output data of the components are available. The existing Matlab-Toolbox used in step 4 (Section 3.2.2) limits AutoMoG to systems that contain components with one independent variable. For example, variable-speed pumps cannot be modeled since the consumed power of a variable-speed pump depends on two independent variables, i.e., its rotational speed and volume flow. However, the method to solve the piecewise-affine regression problems in step 4 is easily replaceable. For this reason, AutoMoG is extended to handle components with more than one independent variable in Chapter 4.

AutoMoG is an easy-to-use method to generate efficient MILP models. Furthermore, AutoMoG is not limited to generating models of multi-energy systems but can generate models of any engineering system. AutoMoG drastically decreases the effort for data-driven model generation, enabling a wider spread of optimization models in real-world applications.

Chapter 4

AutoMoG 3D

In this chapter, we extend the method AutoMoG from modeling a single to multiple independent variables. The resulting AutoMoG 3D method automatically generates MILP optimization models of multi-energy systems that contain components with multiple independent variables. AutoMoG 3D builds on the advantages of the AutoMoG method (Chapter 3), i.e., AutoMoG 3D considers the importance of the components in the context of the overall multi-energy system. AutoMoG 3D is particularly useful for generating efficient MILP optimization models from measured data and, thus, operational optimization. However, AutoMoG 3D is not limited to generating models for operational optimization from measured data but can also generate models for synthesis problems, as shown in Section 4.2. A Python version of the AutoMoG 3D code is also released open-source (Kämper et al., 2021d).

The chapter is structured as follows: In Section 4.1, we describe the general workflow of AutoMoG 3D and discuss in detail the use of hinging-hyperplane trees. In Section 4.2, we apply AutoMoG 3D to model an industrial real-world pump system. In Section 4.3, we conclude with the key findings.

Contents of this chapter have been published in:

Kämper, A., Holtwerth, A., Leenders, L., and Bardow, A. (2021b). AutoMoG 3D: Automated Data-Driven Model Generation of Multi-Energy Systems Using Hinging Hyperplanes. *Frontiers in Energy Research*, 9, 719658.

Contribution report: Principal author, Conceptualization, Methodology, Software, Validation, Formal Analysis, Investigation, Data Curation, Writing - Original Draft, Visualization, Project administration.

4.1 Multidimensional piecewise-affine regression using hinging-hyperplane trees

The AutoMoG method (Chapter 3) is limited to multi-energy systems that contain components with one independent variable. In general, components with multiple independent variables exist in multi-energy systems, e.g., pumps with variable rotational speed and volumetric flow rate. Thus, we extend AutoMoG to AutoMoG 3D, enabling the modeling of multi-energy systems that contain components with multiple independent variables. The general workflow of AutoMoG 3D follows the workflow of AutoMoG that is illustrated in Section 3.2. However, the data-driven modeling of components with multiple independent variables for MILP optimization results in multidimensional piecewise-affine regression problems. AutoMoG 3D uses hinging-hyperplane trees to solve these multidimensional piecewise-affine regressions in step 4 of the workflow (Section 3.2.2) since the original AutoMoG method cannot solve them. In the following, we present the details of the hinging-hyperplane trees we broadly discussed in Chapter 2.2.

Hinging-hyperplane trees are based on the hinging-hyperplane method (Breiman, 1993). The hinging-hyperplanes method generates a hinge function that can be used for regression, classification, and function approximation. The hinge function consists of two hyperplanes and their intersection, called the hinge (Figure 4.1). The resulting hinge function h is continuous. The two hyperplanes h^+ and h^- are described by

$$h^+ = \mathbf{x}^T\boldsymbol{\theta}^+, \ h^- = \mathbf{x}^T\boldsymbol{\theta}^-, \tag{4.1}$$

where $\mathbf{x} = [1, x_1, x_2, ..., x_M]^T$ are the independent variables in M dimensions, and $\boldsymbol{\theta}^+$ and $\boldsymbol{\theta}^-$ are the parameters of the hyperplanes. The hinge $\boldsymbol{\Delta} = \boldsymbol{\theta}^+ - \boldsymbol{\theta}^-$ is the joint of the hyperplanes and fulfills

$$\mathbf{x}^T\boldsymbol{\Delta} = 0. \tag{4.2}$$

The hinge function h is either the minimum or the maximum of the two hyperplanes, depending on the fitting task (cf. Figure 4.1):

$$h = \min \text{ (or max) } \left\{h^+, h^-\right\}. \tag{4.3}$$

Breiman (1993) proposed the hinge-finding algorithm that efficiently determines a good position for the hinge to approximate measured data or a function with two hyperplanes. The hinge-finding algorithm assigns each data point to one hy-

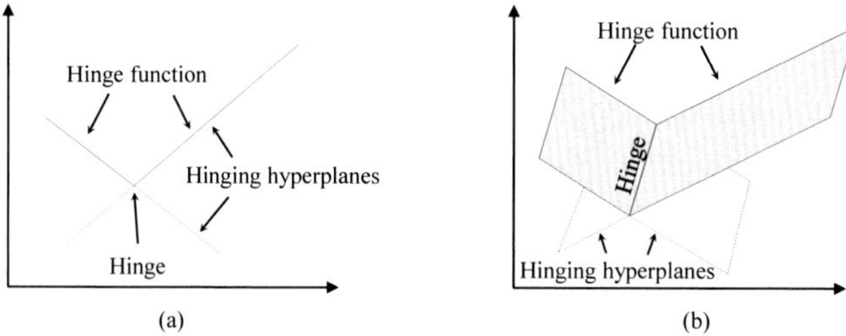

Figure 4.1: Sketch of hinging hyperplanes, the hinge, and the hinge function are illustrated in two dimensions (a) and three dimensions (b), adapted from Pucar and Sjöberg (1998).

perplane and, thus, separates the measured data into 2 data subsets. Ernst (1998) combined the hinge-finding algorithm of Breiman (1993) with binary-tree regression, leading to hinging-hyperplane trees. A hinging-hyperplane tree starts with applying the hinge-finding algorithm to the measured data to be approximated. After the hinge-finding algorithm separated the measured data into 2 data subsets and fitted each data subset with one hyperplane, the worse of the two fitted data subsets is identified. Subsequently, the worst fitted data subset is further refined by applying the hinge-finding algorithm again. By iteratively applying the hinge-finding algorithm, hinging-hyperplane trees can approximate measured data with an arbitrary number of hyperplanes. Ernst (1998) used hinging-hyperplane trees for efficient approximation of nonlinear functions. However, in each iteration, the hinge-finding algorithm only separates the data points of the worst fitted data subset. Thus, in general, hinging-hyperplane trees generate a non-continuous piecewise-affine function.

Hinging-hyperplane trees enable axis-oblique partitioning of measured data. This axis-oblique partitioning allows a more flexible fit to the measured data than axis-orthogonal partitioning used in common tree-based regression (Ernst, 1998).

However, the hinge-finding algorithm of Breiman (1993) that is used in Ernst (1998) suffers from convergence problems (Kenesei and Abonyi, 2015). To solve the convergence problems of the hinge-finding algorithm, Pucar and Sjöberg (1998) showed that the hinge-finding algorithm is equivalent to a Newton algorithm that minimizes the squared residuals between measured data and the hinge function. Pucar and Sjöberg (1998) introduced a damping parameter, following the idea of a damped Newton algorithm, and thereby extended the convergence range of the hinge-finding algorithm. In AutoMoG 3D, we use the improved hinge-finding algorithm (Pucar and Sjöberg, 1998) to calculate hinging-hyperplane trees. The input of the improved hinge-finding algorithm is a data set. The improved hinge-finding algorithm performs a least-squares regression with the fitting criterion V_D that is defined by

$$V_D = \frac{1}{2} \sum_{d=1}^{|D|} (y_d - h(\boldsymbol{x_d}, \boldsymbol{\theta}))^2. \tag{4.4}$$

$|D|$ is the number of data points in the given data set D, y_d, and $\boldsymbol{x_d}$ are the values in data point d from the given data set, and $h(\boldsymbol{x_d}, \boldsymbol{\theta})$ is the value of the hinge function in data point d. The parameter vector $\boldsymbol{\theta}$ contains the parameters for both hyperplanes:

$$\boldsymbol{\theta} = \left(\boldsymbol{\theta}^+, \boldsymbol{\theta}^-\right)^T. \tag{4.5}$$

Thus, the algorithm simultaneously determines the two hyperplanes and thereby the hinge (cf. Equation (4.2)). The improved hinge-finding algorithm applies a damped Newton algorithm to determine the parameter vector $\boldsymbol{\theta}$:

$$\hat{\boldsymbol{\theta}} = \underset{\boldsymbol{\theta}}{\operatorname{argmin}}\ V_D. \tag{4.6}$$

More details of the improved hinge-finding algorithm can be found in Pucar and Sjöberg (1998).

In step 4, AutoMoG 3D uses the hinging-hyperplane trees to add one piecewise-affine region to the model of a component (Figure 4.2). One piecewise-affine region corresponds to one hyperplane.

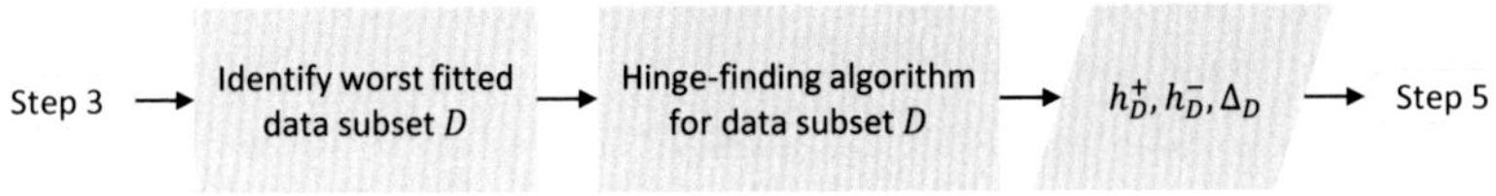

Step 3 → Identify worst fitted data subset D → Hinge-finding algorithm for data subset D → h_D^+, h_D^-, Δ_D → Step 5

Figure 4.2: Flowchart of step 4 in AutoMoG 3D. The worst fitted data subset D is partitioned into two new data subsets D^+ and D^- by the hinge-finding algorithm. The data subsets D^+ and D^- are approximated by the hyperplanes h_D^+ and h_D^-, respectively. The hyperplanes h_D^+ and h_D^- are separated at the hinge Δ_D.

In the following, we exemplify step 4 with an arbitrary component that has been chosen to be refined in step 6 of the previous iteration. The component is already modeled by two piecewise-affine regions (Figure 4.3 (a)). AutoMoG 3D compares the mean squared error in each region of the chosen component. In the given example, the two regions with data subsets D_1^+ and D_1^- exist because the component is modeled by two piecewise-affine regions before applying step 4. The data subset with the highest mean squared error is identified as the worst-fitted data subset D. In this example, the worst-fitted data subset D is D_1^-. D_1^- is refined by the hinge-finding algorithm proposed by Pucar and Sjöberg (1998). The hinge-finding algorithm determines the position of the hinge Δ_{D_2} that partitions the data subset D_1^- into the two new data subsets D_2^+ and D_2^- (Figure 4.3 (b)).

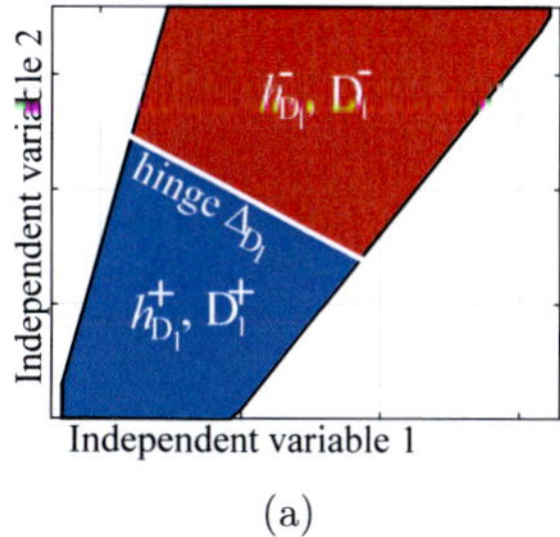

(a)

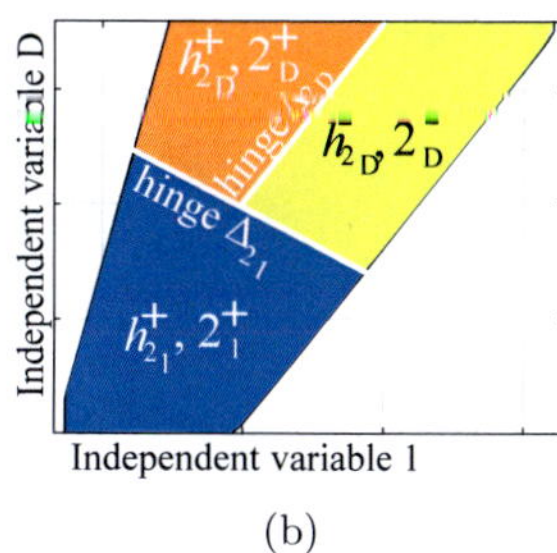

(b)

Figure 4.3: Example of step 4 in AutoMoG 3D for an arbitrary component with two independent variables. On the left (a), the measured data are approximated by two hyperplanes. The hyperplane $h_{\mathrm{D}_1}^+$ approximates the data subset D_1^+ and the hyperplane $h_{\mathrm{D}_1}^-$ approximates the data subset D_1^-. D_1^- is the worst-fitted data subset. Thus, on the right (b), the hyperplane $h_{\mathrm{D}_1}^-$ is partitioned by a new hinge Δ_{D_2} into two new hyperplanes $h_{\mathrm{D}_2}^+$ and $h_{\mathrm{D}_2}^-$ for a better approximation of the measured data.

After the hinge-finding algorithm terminates, the parameters are known for all hyperplanes and hinges of the chosen component. The obtained parameters can directly be used as a component model in an MILP optimization problem.

After step 4, AutoMoG 3D continues with step 5 and checks if the chosen information criterion indicates overfitting for the calculated refinement. When AutoMoG 3D terminates, we obtain an optimization model of the multi-energy system that contains components with multiple independent variables.

4.2 Case Study: Industrial real-world pump system

AutoMoG 3D is applied to model an industrial real-world pump system taken from our previous work (Bahl et al., 2018a). We implemented AutoMoG 3D in Matlab and formulated all subsequent optimization problems in GAMS 27.3.0 (GAMS Development Corporation, 2016).

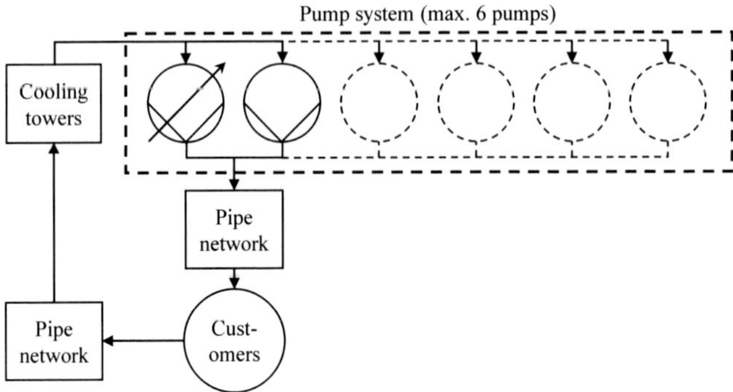

Figure 4.4: Scheme of the industrial real-world pump system (dotted rectangle). The pump system is connected to several customers and cooling towers via a pipe network, adapted from Bahl et al. (2018a).

The scheme of the pump system is shown in Figure 4.4. The purpose of the pump system is to provide cooling water to several customers. The customers differ in their distance to the cooling tower and may have fluctuating cooling-water demands. Since the customer demand changes for different time steps, the pump system has to provide a wide range of volumetric flow rates and pressure differences. Bahl et al. (2018a) and Baumgärtner et al. (2019) solved the synthesis problem for the pump system to reach minimal total annual cost. Thereby, they identified the type and size of each installed pump. At most, six pumps can be installed in the pump system due to a limited power supply. Two pump types with three possible sizes each can be installed: three fixed-speed pumps with nominal volumetric flow rates $\dot{V}^{\mathrm{nom}} = 1000\,\mathrm{m^3\,h^{-1}}, 2000\,\mathrm{m^3\,h^{-1}}$ and $3000\,\mathrm{m^3\,h^{-1}}$, and three variable-speed pumps with the same nominal volumetric flow rates $\dot{V}^{\mathrm{nom}} = 1000\,\mathrm{m^3\,h^{-1}}, 2000\,\mathrm{m^3\,h^{-1}}$ and $3000\,\mathrm{m^3\,h^{-1}}$, and nominal rotational speed $n^{\mathrm{nom}} = 50\,\mathrm{Hz}$. For a fixed-speed pump, the pressure difference and the consumed power are two independent nonlinear functions that depend on its volumetric flow rate $\dot{V}$ only. In contrast, the pressure difference and the consumed power of a variable-speed pump are two independent nonlinear functions that depend on its rotational speed n and volumetric flow rate $\dot{V}$ (Bahl et al., 2018a). The nonlinear functions describe the actual behavior of the pumps. In the following, we refer to these nonlinear functions as the actual nonlinear functions of the pumps. The implementation of the actual nonlinear functions leads to an MINLP optimization model. We use this MINLP optimization problem to benchmark the AutoMoG 3D model.

In this case study, we use AutoMoG 3D to generate two MILP models of the pump system. The first AutoMoG 3D model should reach the same model accuracy as the model used by Bahl et al. (2018a). For this purpose, we set the desired accuracy δ^{rel} in step 3 to the accuracy of the MILP model used by Bahl et al. (2018a) (cf. Figure 3.1). The second AutoMoG 3D model can employ the same model complexity as the model used by Bahl et al. (2018a). For this purpose, we set the stopping criterion in step 3 to the number of piecewise-affine regions to reach a comparable model resolution as the MILP model used by Bahl et al. (2018a). Subsequently, we solve the synthesis problem of the pump system with the generated AutoMoG 3D models and compare our results to the results from Bahl et al. (2018a).

In Section 4.2.1, we employ AutoMoG 3D. In Section 4.2.2, we solve the pump-system synthesis problem and analyze the computational efficiency of the AutoMoG 3D model. In Section 4.2.3, we analyze the solution quality of the AutoMoG 3D model in terms of the total annual cost and the design of the pump system.

4.2.1 Modeling the pump system using AutoMoG 3D

As introduced above, we employ AutoMoG 3D to generate 2 MILP models of the pump system by linearizing the actual nonlinear functions of the pumps. However, since AutoMoG 3D expects data points as input, we sample the nonlinear functions first. We use 2500 data points to sample each actual nonlinear function of a variable-speed pump and 50 data points for each actual nonlinear function of a fixed-speed pump.

In the following, we exemplarily illustrate the model generation for the power function of a variable-speed pump in detail. The consumed power of a variable-speed pump depends on its rotational speed n and volumetric flow rate $\dot{V}$, i.e., on two independent variables. The actual nonlinear power function of the variable-speed pump is given by a third-degree polynomial function (Figure 4.5 (a)).

To obtain an MILP optimization model of the pump system, we apply AutoMoG 3D. For the variable-speed pump, AutoMoG 3D needs four piecewise-affine regions to reach the accuracy of the model used by Bahl et al. (2018a) (Figure 4.5 (b)). We refer to the corresponding model as AutoMoG 3D model (accuracy) in the following. To reach a comparable resolution as the model used by Bahl et al. (2018a), AutoMoG 3D needs 18 piecewise-affine regions (Figure 4.5 (c)). We refer to the corresponding model as AutoMoG 3D model (resolution) in the following. Here, we define comparable resolution by the average area of one piecewise-affine region in the input space.

Bahl et al. (2018a) manually linearized the actual nonlinear power function with the generalized-convex-combination method (GCCM) (Geißler, 2011; Geißler et al., 2012) to obtain an MILP optimization model. The generalized-convex-combination method cannot consider the actual operating limits of the power function but needs an axis-orthogonal, rectangular grid. As a result, Bahl et al. (2018a) used 40 piecewise affine regions to model the actual nonlinear power function (Figure 4.5 (d)). In contrast, AutoMoG 3D automatically considers the actual operating limits for the power function of the variable-speed pump by the convex hull (cf. step 1 in Figure 3.1).

To compare the model accuracy, we show the relative root squared error of the power function $\epsilon^{\text{power}}(n, \dot{V})$ over the entire input space (Figure 4.6). The relative root squared error of the power function ϵ^{power} is defined as

$$\epsilon^{\text{power}}(n, \dot{V}) = \frac{\sqrt{\left(P^{\text{real}}(n, \dot{V}) - P^{\text{model}}(n, \dot{V})\right)^2}}{P^{\text{real}}(n, \dot{V})} , \tag{4.7}$$

with the actual power P^{real} and the modeled power P^{model}.

The AutoMoG 3D model (accuracy) and the generalized-convex-combination model show a small relative error $\epsilon^{\text{power}}(n, \dot{V})$ over a wide operating range (Figure 4.6 (a) and (c)). Both models have an average relative error $\overline{\epsilon^{\text{power}}}$ of 0.01. However, the AutoMoG 3D model (accuracy) uses significantly fewer piecewise-affine regions to approximate the actual nonlinear power function of the variable-speed pump: 4 vs. 40. For both models, the relative error $\epsilon^{\text{power}}(n, \dot{V})$ increases for low rotational speed n and low volumetric flow rate $\dot{V}$ and, in particular, on the edges of the piecewise-affine regions. The error increases for low rotational speed n and low volumetric flow rate $\dot{V}$ because the actual nonlinear function is more curved in this area. Furthermore, the error increases on the edges of the piecewise-affine regions since the hinge-finding algorithm minimizes the mean squared error and, thus, tends to a worse fit on the edges.

The AutoMoG 3D model (resolution) has an average relative error $\overline{\epsilon^{\text{power}}}$ of 0.001, and the generalized-convex-combination model has an average relative error $\overline{\epsilon^{\text{power}}}$ of 0.01 (Figure 4.6 (b) and (c)). However, both models have the same resolution. The AutoMoG 3D model (resolution) reaches a smaller relative error ϵ^{power} because

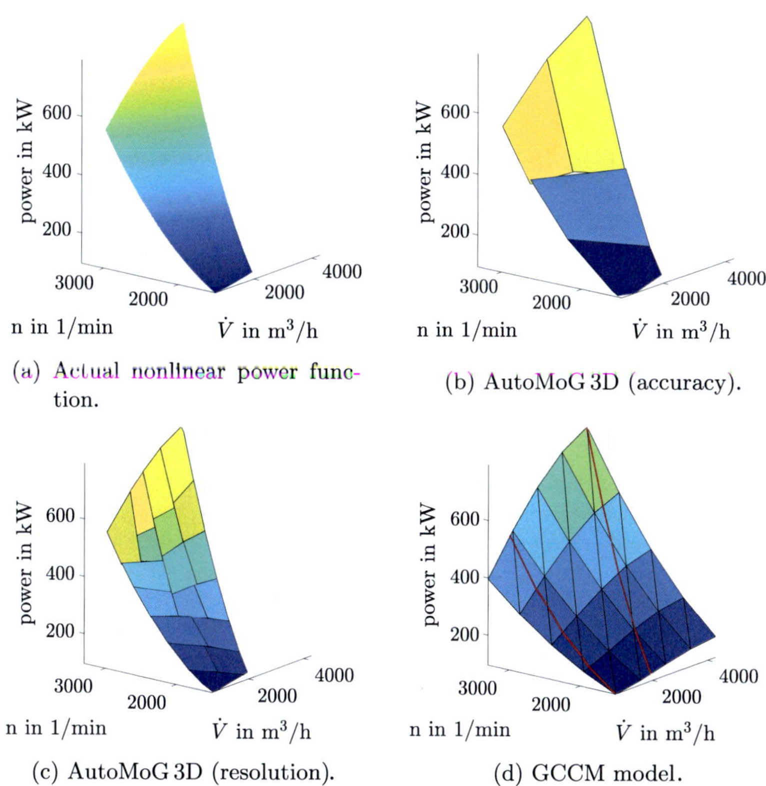

(a) Actual nonlinear power function.

(b) AutoMoG 3D (accuracy).

(c) AutoMoG 3D (resolution).

(d) GCCM model.

Figure 4.5: Actual nonlinear power function of a variable-speed pump (a), AutoMoG 3D model with the same accuracy as the model of the generalized-convex-combination method (GCCM model) (b), AutoMoG 3D model with a comparable resolution as the GCCM model (c), and the GCCM model used by Bahl et al. (2018a) (d). The power of the modeled pump depends on the rotational speed n and the volumetric flow rate $\dot{V}$. The generalized-convex-combination method needs a rectangular grid, whereas AutoMoG 3D automatically considers the actual operating limits of the pump. The red lines in (d) show the actual operating limits. The color code only supports the visual differentiation of the piecewise-affine regions.

AutoMoG 3D refines the regions of the current worst fit. In contrast, the generalized-convex-combination method is based on an inflexible grid.

In summary, compared to the generalized-convex-combination method, AutoMoG 3D can

1. generate a model with comparable resolution but ten times higher accuracy.
2. generate a model with ten times lower resolution but the same accuracy.

For completeness, we briefly show the results of the model generation for the pressure difference of the same variable-speed pump (Figure 4.7). The pressure head is shown instead of the pressure difference because the pump manufacturer used the head to describe the pressure difference of the variable-speed pump. The head is the height of a liquid column corresponding to the pressure exerted by this liquid column on its bottom.

The AutoMoG 3D model (accuracy) and the generalized-convex-combination model have an average relative error $\overline{\epsilon^{\text{head}}}$ of 0.007 (Figure 4.7 (b) and (d)). However, the AutoMoG 3D model (accuracy) again uses only four piecewise-affine regions to approximate the actual nonlinear head function of the variable-speed pump. The generalized-convex-combination model reaches a smaller average relative error $\overline{\epsilon}$ for the pressure head than for the power because the inflexible GCCM grid turns out to fit the nonlinear head function better than the nonlinear power function. However, this behavior cannot be guaranteed in general. The AutoMoG 3D model (accuracy) reaches the same average relative error $\overline{\epsilon^{\text{head}}}$ as the generalized-convex-combination model but with fewer regions (4 vs. 40) due to the flexibility of the hinging-hyperplane trees. Furthermore, the AutoMoG 3D model (resolution) again has the lowest average relative error $\overline{\epsilon^{\text{head}}}$ of 0.001 (Figure 4.7 (c)). The generalized-convex-combination and AutoMoG 3D models use the same number of piecewise-affine regions to model the head and the power function. However, the hinging-hyperplane trees choose the positions of the hinges based on the respective data points such that the two different functions of head and power are fitted with high accuracy.

Concerning computational cost, AutoMoG 3D generates the models of all six pumps within 1 min.

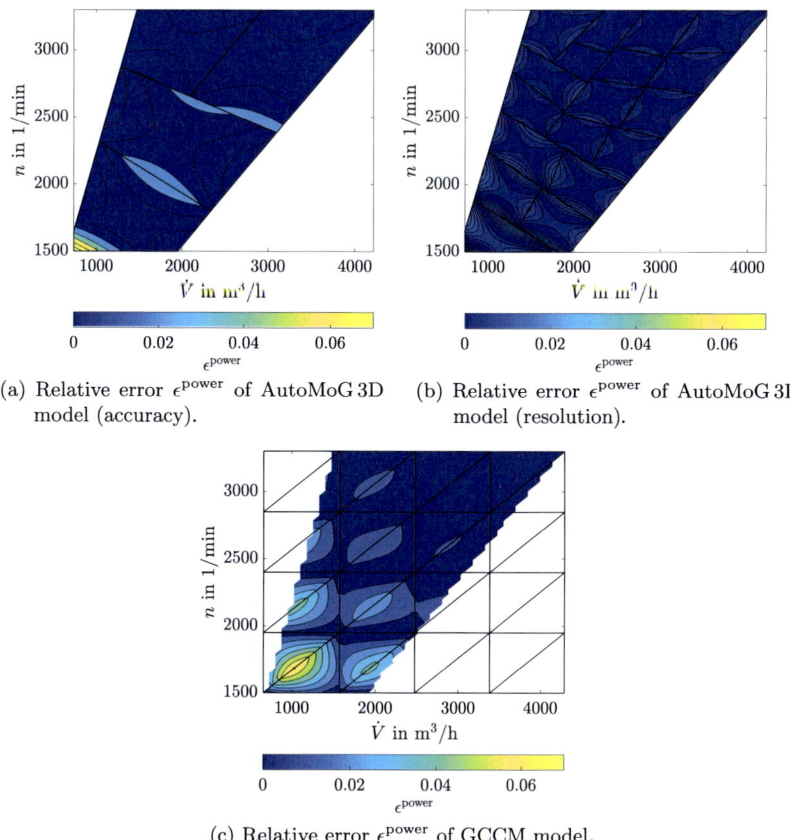

(a) Relative error ϵ^{power} of AutoMoG 3D model (accuracy).

(b) Relative error ϵ^{power} of AutoMoG 3D model (resolution).

(c) Relative error ϵ^{power} of GCCM model.

Figure 4.6: Relative error of the AutoMoG 3D model with same accuracy (a) and comparable resolution (b) compared to the relative error of the generalized-convex-combination method (GCCM) model (c) for the power function of a variable-speed pump.

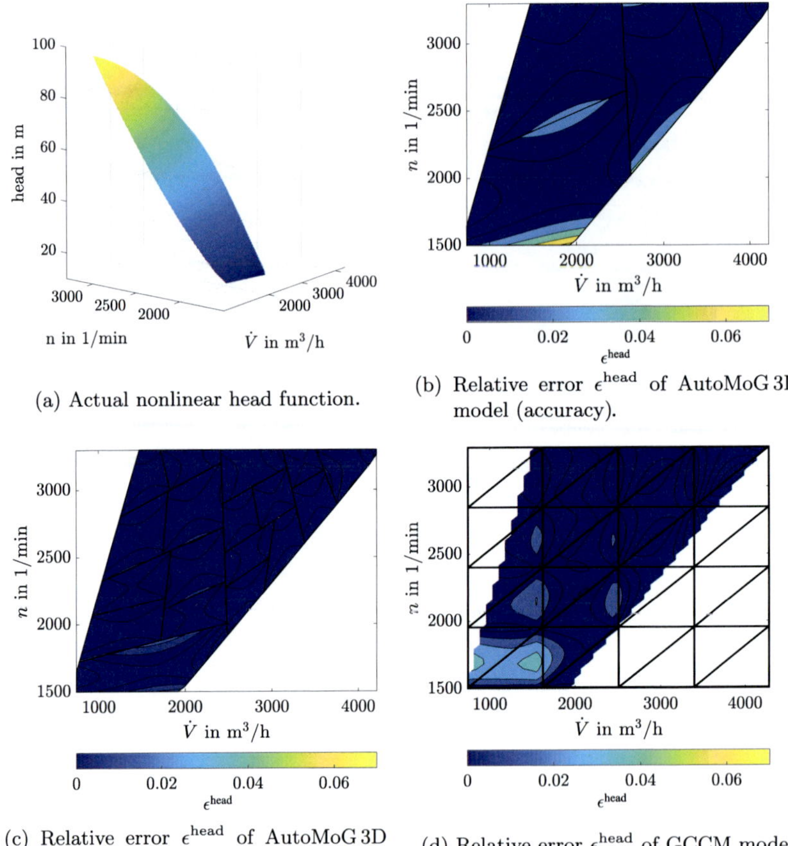

(a) Actual nonlinear head function.

(b) Relative error ϵ^{head} of AutoMoG 3D model (accuracy).

(c) Relative error ϵ^{head} of AutoMoG 3D model (resolution).

(d) Relative error ϵ^{head} of GCCM model.

Figure 4.7: Actual nonlinear head function of a variable-speed pump (a), and the relative errors of the AutoMoG 3D model with the same accuracy (b) and comparable resolution (c) compared to the relative error of the generalized-convex-combination method (GCCM) model (d) for the head function of a variable-speed pump.

4.2.2 Computational efficiency of the AutoMoG 3D model

In this section, the computational efficiency of the AutoMoG 3D models is compared to two other models of the pump system. For this purpose, we solve the pump-system synthesis problem with four models: 1) the AutoMoG 3D model with the same accuracy as the generalized-convex-combination model, 2) the AutoMoG 3D model with comparable resolution as the generalized-convex-combination model, 3) the generalized-convex-combination model, and 4) the MINLP model with the actual nonlinear functions of all pumps. All optimization problems are solved using four Intel-Xeon CPUs at 3.2 GHz and 25 GB RAM. We solve all MILPs using CPLEX 12.9.0.0 (IBM Corporation, 2017) and all MINLPs using BARON 19.3.24 (Zhou et al., 2018) with a time limit of 48 h and an optimality gap of 2 %

The pump-system synthesis problem uses an original time series of the cooling-water demand and the pressure difference consisting of two years with a time-step length of 1 d. We generate five instances of the original time series with variations of ±5 % using latin-hypercube sampling (McKay et al., 2000). We aggregate the time series to seven representative time steps with the method of Bahl et al. (2018a).

The computation times of the synthesis problems are shown in Figure 4.8. On average, the AutoMoG 3D model with the same accuracy as the generalized-convex-combination model solves the pump-system synthesis problem ten times faster than the generalized-convex-combination model (290 s vs. 2897 s). The computation time is significantly shorter since the AutoMoG 3D model has significantly fewer piecewise-affine regions (e.g., 4 for the power function of a variable-speed pump) than the generalized-convex-combination model (e.g., 40 for the power function of a variable-speed pump). The fewer piecewise-affine regions result in fewer binary variables and, thus, a computationally more efficient optimization model. After preprocessing, the AutoMoG 3D model has 2800 binary variables for the pump-system synthesis problem, whereas the generalized-convex-combination model has 8200 binary variables.

The AutoMoG 3D model with comparable resolution as the generalized-convex-combination model is faster for four out of five instances and solves the pump-system synthesis problem 23 % faster than the generalized-convex-combination model on average (2224 s vs. 2897 s, Figure 4.8). The AutoMoG 3D model has 5500 binary variables for the pump-system synthesis problem after preprocessing and thus still 33 % fewer binary variables. Still, the AutoMoG 3D model has higher accuracy with a relative model error $\overline{\epsilon^{\text{power}}}$ of 0.1 % compared to 1.0 % in the generalized-convex-combination model (cf. Section 4.2.1).

The MINLP model leads to optimality gaps between 85 % and 90 % at the time limit of 48 h. These optimality gaps show that the MINLP model is computationally much more demanding than the MILP models in the case study.

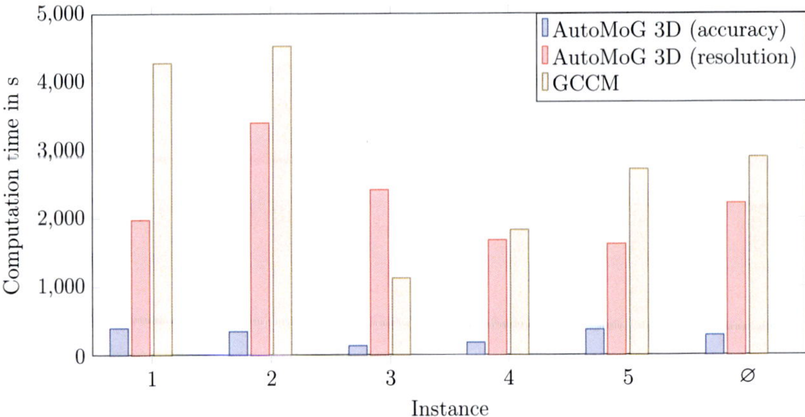

Figure 4.8: Computation time to satisfy the optimality gap of 2 % for all instances and all models except the MINLP model. The MINLP model did not fulfill the optimality gap in any instance at the time limit of 48 h. Therefore, there are no computation times for the MINLP model.

4.2.3 Solution quality of the AutoMoG 3D model

After comparing the computational efficiency, we compare the solution quality of the same four models by analyzing the objective of the pump-system synthesis problem and the resulting design of the pump system.

4.2.3.1 Objective of the pump-system synthesis problem

The pump-system synthesis problem minimizes the total annual cost. To compare the solution quality of the four models, we calculate their actual total annual cost. We define the actual total annual cost as the cost that occurs if we fix the solution of a model and recalculate the total annual cost of this solution in the MINLP model.

The synthesis problems provide the chosen pumps, their sizes, and their operating points. We fix the pumps, their sizes, and their operating status for each time step in the MINLP model. Then, we recalculate the MINLP model with the actual nonlinear power functions to obtain the actual total annual cost of each model and instance (Figure 4.9).

For each instance, the 3 MILP solutions have lower actual total annual costs than the best feasible MINLP solution at the time limit of 48 h. In the following, we compare the MILP solutions with each other. The actual total annual costs of the three MILP models do not differ more than 1 % for any instance. None of the MILP models provides solutions with a lower actual total annual cost for all instances. Thus, none of the MILP models can be claimed to systematically identify solutions with lower actual total annual cost than the other MILP models. All MILP models show a comparable solution quality (Figure 4.9). However, the solutions of the AutoMoG 3D models are found faster (Figure 4.8). To demonstrate the need for accurate models, we also modeled all pumps with only one affine region and solved the synthesis problem with these models. In that case, the actual total annual cost increases by 10 % to 2.62×10^6 € on average. This result shows that more accurate models like the former MILP models are necessary for a high solution quality.

In summary, AutoMoG 3D automatically generates pump-system models in 1 min and provides computationally more efficient models than the generalized-convex-combination method. Still, the solution quality of the AutoMoG 3D models is as high as the solution quality of the generalized-convex-combination model.

4.2.3.2 Design of the pump-system

Table 4.1 shows the resulting pump system design for all instances and models. The designs resulting from the three MILP models consist of more fixed-speed pumps than variable-speed pumps in almost every instance. At least one variable-speed pump is chosen for each instance. This result seems meaningful: Fixed-speed pumps operate more efficiently at nominal volumetric flow rate than variable-speed pumps since variable-speed pumps need frequency converters that cause efficiency losses. At the same time, fixed-speed pumps operate inefficiently at variable volumetric flow rates since the volumetric flow rate is reduced by throttling the pressure difference. Thus, the pump system operates such that the variable-speed pumps balance the fluctuations of the volumetric-flow demand. The variable-speed pumps allow the fixed-speed pumps to operate efficiently at their nominal volumetric flow rate.

The designs from the MINLP model consist of more variable-speed pumps than fixed-speed pumps in every instance. These solutions of the MINLP are feasible but lead to higher total annual costs (Figure 4.9).

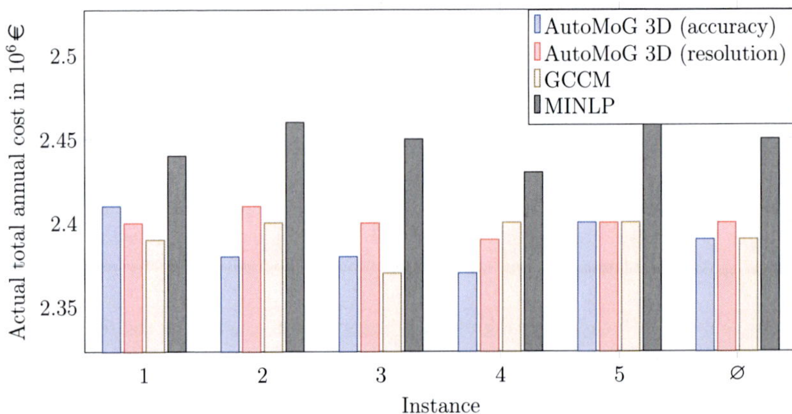

Figure 4.9: Actual total annual cost for all instances and models. The actual total annual cost is determined by fixing the solution of every model and recalculating the total annual cost of this solution in the MINLP model. All MILP models find a better solution in all instances than the MINLP model finds within 48 h.

Table 4.1: Design of the pump system for all instances and models. The table lists the nominal volumetric flow rate of each installed pump. The nominal volumetric flow rate $\dot{V}^{\mathrm{nom}}$ of each installed pump is presented in $1000\,\mathrm{m}^3\,\mathrm{h}^{-1}$.

Instance	Pump type	AutoMoG 3D (accuracy)	AutoMoG 3D (resolution)	GCCM	MINLP
1	Fixed-speed	3, 3, 2	3, 2, 2	3, 3, 2	2, 2,
	Variable-speed	3	2, 2	3	3, 2, 1
2	Fixed-speed	3, 2, 2, 2	3, 3, 2	3, 2, 2, 2, 1	-
	Variable-speed	2	3	2	3, 3, 2, 2
3	Fixed-speed	3, 3, 2, 1	3, 3	3, 2, 2, 2	2, 1
	Variable-speed	2	3, 2	2	2, 2, 2, 1
4	Fixed-speed	3, 2, 2, 2	3, 2, 2, 1	3, 3, 2	2, 2
	Variable-speed	2	2, 1	3	2, 2, 2
5	Fixed-speed	3, 2, 2	3, 3	3, 2, 2, 2, 1	3
	Variable-speed	2, 2	3, 2	2	3, 3, 1

4.3 Conclusion

The AutoMoG 3D method is proposed for multidimensional automated data-driven model generation of multi-energy systems. AutoMoG 3D generates MILP optimization models from measured data of multi-energy systems. AutoMoG 3D can model components with multiple independent variables using hinging-hyperplane trees. Thus, AutoMoG 3D overcomes the main limitation of the original AutoMoG method, which was limited to systems containing components with only one independent variable. However, in general, hinging-hyperplane trees do not generate continuous functions of the component models. To overcome this drawback, other methods to solve the multidimensional piecewise-affine regression problems can easily be used in AutoMoG 3D.

In the case study, a real world pump system is modeled. AutoMoG 3D needs significantly fewer piecewise-affine regions to approximate the actual functions of the pumps with comparable accuracy as the generalized-convex-combination model. AutoMoG 3D provides the pump-system models within 1 min.

The computational efficiency and accuracy of the AutoMoG 3D models were shown by solving the pump-system synthesis problem. The AutoMoG 3D model solves ten times faster than the generalized-convex-combination model. Still, the solution quality of the AutoMoG 3D model is the same as for the generalized-convex-combination model. The performance of the MINLP model is much worse, with an average optimality gap of 88.5 % at the time limit of 48 h, whereas AutoMoG 3D found a better solution in 5 min. Thus, AutoMoG 3D generates an accurate and computationally efficient model of the industrial real-world pump system.

The AutoMoG 3D method can be applied either if measured input and output data of the components are available or if the actual functions of the components are available. AutoMoG 3D drastically reduces the effort for both data-driven modeling and optimization. Thereby, AutoMoG 3D empowers broader applicability of MILP optimization models in real-world applications.

Part II

Accelerating Operational Optimization of Multi-Energy Systems

Chapter 5

Adaptive Rolling Horizon for Operational Optimization of Multi-Energy Systems

In this chapter, we propose the Adaptive-Rolling-Horizon approach to accelerate the operational optimization of multi-energy systems. Adaptive Rolling Horizon copes with the second major task that we identified in Chapter 2.4: finding high-quality solutions for operational optimizations of multi-energy systems in a short time. Adaptive Rolling Horizon uses a fixed step size $\mathrm{T}^{\mathrm{step}}$ and adaptively determines the foresight $\mathrm{T}_k^{\mathrm{fore}}$ for each subproblem k by employing knowledge about the multi-energy system. Thereby, we accelerate operational optimization and reach a high solution quality. We show that a fixed foresight $\mathrm{T}^{\mathrm{fore}}$ can lead to unnecessarily high computation times or low solution quality.

In Section 5.1, we state an MILP formulation for operational optimization of multi-energy systems. In Section 5.2, we describe the proposed Adaptive Rolling Horizon. In Section 5.3, we apply the Adaptive Rolling Horizon to a real-world case study. In Section 5.4, we conclude with the key findings.

Major parts of this chapter are reproduced by permission of ECOS 2021 Organizing Committee from:

Kämper, A., Geers, P., Leenders, L., and Bardow, A. (2021a). Adaptive Rolling Horizon for operational optimization of multi-energy systems. In *Proceedings of the ECOS 2021 - 34th International Conference on Efficiency, Cost, Optimization, Simulation and Environmental Impact of Energy Systems (ECOS 2021).*

Contribution report: Principal author, Conceptualization, Methodology, Software, Validation, Formal Analysis, Investigation, Data Curation, Writing - Original Draft, Visualization, Project administration.

5.1 MILP operational optimization problem of multi-energy systems

Before we describe the Adaptive-Rolling-Horizon approach, we formulate a generic MILP operational optimization problem of multi-energy systems. In MILP operational optimization, the MILP models provided by the methods from Chapters 3 and 4 can easily be used. However, the MILP operational optimization problem can be extended by further constraints, e.g., minimum up- and downtimes, ramping constraints, or start-up costs. The objective of operational optimization is typically to minimize the operational expenditures $OPEX$ (Equation (5.1)):

$$\min \quad OPEX = \sum_{t \in T} \left(\sum_{e \in E} \left(\overbrace{O^{\text{source}}_{e,t} c^{\text{buy}}_{e,t}}^{\textit{Energy purchase}} - \overbrace{I^{\text{sink}}_{e,t} c^{\text{sell}}_{e,t}}^{\textit{Energy sales}} \right) + \overbrace{\sum_{s \in s} \delta^{\text{start}}_{s,t} c^{\text{start}}_{s}}^{\textit{Start-up costs}} \right). \quad (5.1)$$

The operational expenditures $OPEX$ consist of the costs for purchasing energy, revenues for selling energy, and start-up costs for components of the multi-energy system. The constraints of the generic MILP operational optimization problem are introduced in Equations (5.2) to (5.14).

$$\begin{aligned}
d_{e,t} &= \sum_{s \in S} O_{s,e,t} - \sum_{s \in S^{\text{input}}_{e}} I_{s,t} \\
&\quad + O^{\text{source}}_{e,t} p^{\text{source}}_{e} - I^{\text{sink}}_{e,t} p^{\text{sink}}_{e}, && \forall e \in E, t \in T, && (5.2)\\
I_{s,t} &= \sum_{n \in N_{s,e}} \left(\hat{\delta}_{s,e,t,n} b_{s,e,n} + \hat{O}_{s,e,t,n} a_{s,e,n} \right), && \forall s \in S, e \in E, t \in T, && (5.3)\\
\hat{O}_{s,e,t,n} &\leq \hat{\delta}_{s,e,t,n} o^{\text{ub}}_{s,e,t,n}, && \forall s \in S, t \in T, n \in N_{s,e}, && (5.4)\\
\hat{O}_{s,e,t,n} &\geq \hat{\delta}_{s,e,t,n} o^{\text{ub}}_{s,e,t,n-1}, && \forall s \in S, t \in T, n \in N_{s,e} \setminus \{0\}, && (5.5)\\
O_{s,e,t} &= \sum_{n \in N_{s,e}} \hat{O}_{s,e,t,n}, && \forall s \in S, e \in E, t \in T, && (5.6)\\
\delta_{s,t} &= \sum_{n \in N_{s,e}} \hat{\delta}_{s,e,t,n}, && \forall s \in S, e \in E, t \in T, && (5.7)\\
0 &= \delta_{s,t-1} - \delta_{s,t} - \delta^{\text{shutdown}}_{s,t} + \delta^{\text{start}}_{s,t}, && \forall s \in S, t \in T \setminus \{0\}, && (5.8)\\
1 &\geq \delta^{\text{shutdown}}_{s,t} + \delta^{\text{start}}_{s,t}, && \forall s \in S, t \in T, && (5.9)
\end{aligned}$$

$$\begin{aligned}
\delta_{s,t}^{\text{start}} &\leq \delta_{s,\tau}, \quad \forall s \in S, t \in T, \tau \in \{\tau \mid t \leq \tau \leq t + T_s^{\text{min,up}}\}, && (5.10)\\
1 - \delta_{s,t}^{\text{shutdown}} &\geq \delta_{s,\tau}, \quad \forall s \in S, t \in T, \tau \in \{\tau \mid t \leq \tau \leq t + T_s^{\text{min,down}}\}, && (5.11)\\
I_{s,t} - I_{s,t-1} &\leq r_s^{\text{up}} \delta_{s,t-1} + m_s^{\text{up}} \cdot (1 - \delta_{s,t-1}), \quad \forall s \in S, t \in T \setminus \{0\}, && (5.12)\\
I_{s,t-1} - I_{s,t} &\leq r_s^{\text{down}} \delta_{s,t} + m_s^{\text{down}} \cdot (1 - \delta_{s,t}), \quad \forall s \in S, t \in T \setminus \{0\}, && (5.13)\\
\boldsymbol{b} &\geq \boldsymbol{A} \boldsymbol{z}^{\text{T}}. && (5.14)
\end{aligned}$$

Equations (5.2) represent the energy balances. The demand $d_{e,t}$ of each energy form e has to be fulfilled at every time step t. $O_{s,e,t}$ and $I_{s,e,t}$ are output and input of component s for energy form e at time step t. S_e^{input} is the set of components using energy form e as input. The binary parameters p_e^{source} and p_e^{sink} indicate whether energy form e has a source or a sink. $O_{e,t}^{\text{source}}$ is the energy purchased from a source (e.g., gas supply), and $I_{e,t}^{\text{sink}}$ is energy sold to a sink (e.g., electricity to the grid). Equations (5.3) to (5.7) represent the piecewise-affine input-output relationship of the component s (cf. Equations (3.1) to (3.5)). The parameters $a_{s,e,n}$ and $b_{s,e,n}$ denote the slope and the intercept of linear section n. The binary variable $\hat{\delta}_{s,e,t,n}$ is equal to one if and only if the output $\hat{O}_{s,e,t,n}$ lies between the upper bound $o_{s,e,n}^{\text{ub}}$ of the linear section n (Equation (5.4)) and the upper bound $o_{s,e,n-1}^{\text{ub}}$ of the lower linear section $n-1$ (Equation (5.5)). Equations (5.7) ensure that only one linear section n for energy form e is active when component s is active at time step t. Equations (5.8) detect component start-ups, and Equations (5.9) ensure that the components cannot start up and shut down at the same time step t. Equations (5.10) and (5.11) ensure minimum up and downtimes $T_s^{\text{min,up}}$ and $T_s^{\text{min,down}}$ of the components. Equations (5.12) and (5.13) are ramping constraints. The ramping parameters r_s^{up} and r_s^{down} limit the maximum load increase and decrease of component s between two time steps. The constraints use a Big-M formulation to enable ignoring the ramping constraints during component start-ups and shut-downs. The Big-M value m_s^{up} denotes the maximum start-up ramp, and m_s^{down} denotes the maximum shut-down ramp of component s. Equation (5.14) is generic and represents further constraints that can occur in MILP operational optimization of multi-energy systems, e.g., minimum and maximum component loads. $\boldsymbol{z}$ is a vector that incorporates all variables of the optimization problem.

Directly solving this operational optimization problem (Equations (5.1) to (5.14)) often requires high computation times. In practice, strict time limits prohibit such high computation times. Thus, we propose the Adaptive-Rolling-Horizon approach to accelerate the operational optimization of multi-energy systems.

5.2 General workflow of Adaptive Rolling Horizon

The general workflow of the Adaptive Rolling Horizon is based on the general workflow of the classic Rolling Horizon (Figure 2.1). The Adaptive Rolling Horizon aims to reduce total computation times while retaining high solution quality. Marquant et al. (2015) show that the step size T^{step} has a high impact on the computation time, and the foresight T^{fore} has a high impact on the solution quality. Thus, Adaptive Rolling Horizon determines a step size T^{step} that allows for short computation times and adaptively determines the foresight $\mathrm{T}_k^{\text{fore}}$ for a high solution quality. $\mathrm{T}_k^{\text{fore}}$ is identified for each Rolling-Horizon subproblem k based on knowledge of the system to be optimized.

The general idea of Adaptive Rolling Horizon can be applied to any problem where Rolling Horizon is used. Here, we present the concept of the Adaptive-Rolling-Horizon approach for a short-term operational optimization of multi-energy systems (Figure 5.1):

Step 1: Adaptive Rolling Horizon determines the step size T^{step} once for the considered system to decompose the original optimization problem into smaller subproblems.

Step 2: Adaptive Rolling Horizon determines a suitable foresight $\mathrm{T}_k^{\text{fore}}$ for subproblem k using information about time-varying and time-coupling effects within the multi-energy system, e.g., current energy prices and the components' operating states. Determining the foresight $\mathrm{T}_k^{\text{fore}}$ is the main innovation of Adaptive Rolling Horizon.

Step 3: Adaptive Rolling Horizon solves the current subproblem k consisting of $(\mathrm{T}^{\text{step}} + \mathrm{T}_k^{\text{fore}})$ time steps.

Step 4: Adaptive Rolling Horizon checks if the end of the full time horizon is reached. If yes, the solution for all time steps in subproblem k is fixed (step size T^{step} and foresight $\mathrm{T}_k^{\text{fore}}$), and the Adaptive-Rolling-Horizon approach terminates. If not, the Adaptive-Rolling-Horizon approach continues with step 5.

Step 5: Adaptive Rolling Horizon fixes the solution for all time steps in the step size T^{step} of subproblem k and saves the operating states of all components at the beginning of the subsequent subproblem $k{+}1$. The Adaptive-Rolling-Horizon approach continues with step 2.

In the following, we describe how to determine the step size T^{step} (step 1) and the foresight $\mathrm{T}_k^{\text{fore}}$ (step 2).

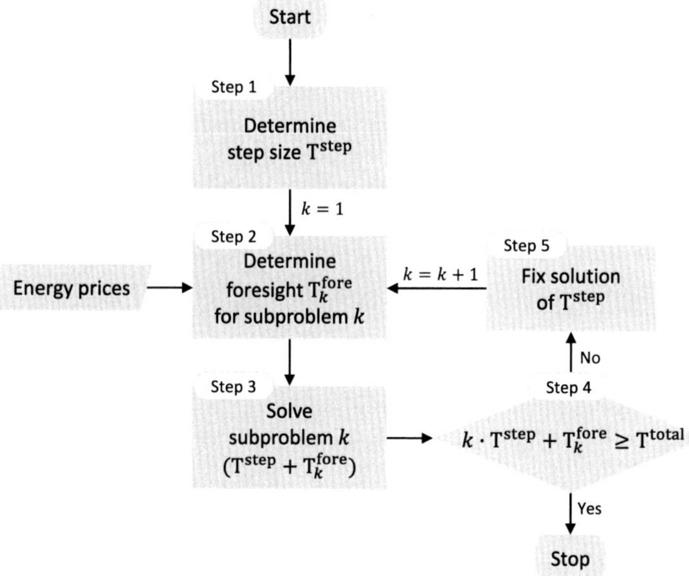

Figure 5.1: Flowchart of the Adaptive-Rolling-Horizon approach.

5.2.1 Step 1: Determination of the step size T^{step}

The step size T^{step} has a high impact on computation time. The step size T^{step} for a minimal total computation time results from a trade-off between the number of subproblems and the length of each subproblem (Marquant et al., 2015). A short step size T^{step} leads to smaller but many subproblems and, thus, a high total computation time. A long step size T^{step} leads to few but large subproblems with high computation times and, thus, also to a high total computation time.

To find a suitable value for the step size T^{step}, we use a brute-force approach. We use historical data from the energy system and perform operational optimizations with the desired full time horizon. In the operational optimizations, we vary the step size T^{step} and the foresight T^{fore} for time series from all seasons (Figure 5.2). For the investigated multi-energy system, the time series are the energy demands and prices. The step size T^{step} can be increased until the total computation time strongly increases.

Figure 5.2 shows the trade-off in computation time for the step size $\mathrm{T}^{\mathrm{step}}$. The original optimization problem has a full time horizon of four weeks with an hourly resolution (672 time steps). A small step size $\mathrm{T}^{\mathrm{step}}$ results in many small subproblems and, thus, high total computation time. A large step size $\mathrm{T}^{\mathrm{step}}$ results in a few large subproblems and, thus, high total computation time, too. In between, we identify a broad interval in which the average total computation time is short (approx. between $\mathrm{T}^{\mathrm{step}}$= 16 time steps and $\mathrm{T}^{\mathrm{step}}$= 32 time steps).

From this interval, the user chooses the fixed step size $\mathrm{T}^{\mathrm{step}}$. Choosing a short step size $\mathrm{T}^{\mathrm{step}}$ is a particularly good idea when considering minimum up- and downtimes of components. A shorter step size $\mathrm{T}^{\mathrm{step}}$ leads to subproblems where start-ups or shut-downs can be excluded when the minimum up- and downtimes exceed the step size. Excluding start-ups or shut-downs in a subproblem reduces computation time.

For the investigated multi-energy system, the step size $\mathrm{T}^{\mathrm{step}}$ for minimal computation times does not strongly depend on the foresight $\mathrm{T}^{\mathrm{fore}}$ and the time series of energy demands and prices (Figure 5.2). Thus, we assume that the step size $\mathrm{T}^{\mathrm{step}}$ for minimal computation times does not change when changing the foresight $\mathrm{T}_k^{\mathrm{fore}}$. Based on this assumption, we choose a fixed step size $\mathrm{T}^{\mathrm{step}}$ while determining a specific foresight $\mathrm{T}_k^{\mathrm{fore}}$ for each subproblem k to reach high solution quality. However, this assumption should be checked when Adaptive Rolling Horizon is applied to another multi-energy system.

5.2.2 Step 2: Determination of the foresight $\mathrm{T}^{\mathrm{fore}}$

The Adaptive Rolling Horizon aims to determine a foresight $\mathrm{T}_k^{\mathrm{fore}}$ for each subproblem k as short as possible to reach short total computation times but as long as necessary to retain high solution quality. The foresight $\mathrm{T}_k^{\mathrm{fore}}$ has a high impact on solution quality (Marquant et al., 2015). A foresight $\mathrm{T}_k^{\mathrm{fore}}$ is suitable for subproblem k if it captures the key events influencing the objective function such that the solver makes the best decisions in terms of the objective for the time steps in the step size $\mathrm{T}_k^{\mathrm{step}}$.

The objective function of operational optimization is typically the operational expenditure ($OPEX$). The full time horizon is often in the order of weeks to capture optimal start-ups and shut-downs of the components in the system. Optimal start-ups and shut-downs are crucial for the $OPEX$ due to the start-up costs of the components.

Applying a common Rolling Horizon may lead to suboptimal start-up and shut-down decisions. As an extreme example, a component might generate a constant profit of 500 €/24 h. The start-up cost of the component is 600 €. If we consider subproblems with 24 h time horizon, the component is never switched on, which is

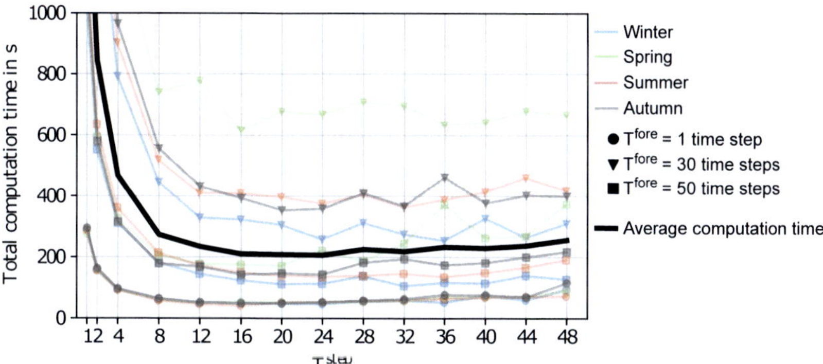

Figure 5.2: Total computation time over step size T^{step} for three foresights T^{fore} (1, 30, and 50 time steps) and time series from four seasons (blue, green, red, and grey) for the exemplary multi-energy system investigated in the case study. The average computation time is shown in black. The total computation time is minimal between $T^{step} = 16$ time steps and $T^{step} = 32$ time steps.

obviously the wrong decision. Adaptive Rolling Horizon aims to determine the minimum foresight T_k^{fore} for each subproblem k while ensuring correct start-ups of the components. This strong requirement leads to a complex case distinction to determine the foresight T_k^{fore} as shown in the following. Simpler algorithms to determine the foresight T_k^{fore} might be possible. However, we believe that simpler algorithms either lead to wrong start-ups of components or unnecessarily long foresight.

A sufficiently long foresight T_k^{fore} is needed such that the profit in subproblem k can exceed the start-up cost. Thus, the Adaptive-Rolling-Horizon approach aims to identify the break-even time where the profits compensate the start-up cost. From this break-even time, the Adaptive-Rolling-Horizon approach determines the foresight T_k^{fore} for each subproblem k such that subproblem k contains sufficient information to make optimal start-up and shut-down decisions. Note that the Adaptive Rolling Horizon is a heuristic approach and does not guarantee the globally optimal solution of the original optimization problem.

In each subproblem k, the Adaptive Rolling Horizon calculates a suitable foresight for each component $T_{k,s}^{fore}$, and chooses the maximum value as foresight

$$T_k^{fore} = \max_{S} T_{k,s}^{fore} \,. \tag{5.15}$$

Determining a suitable foresight $\mathrm{T}^{\text{fore}}_{k,s}$ relies on, on the one hand, the current state of component s (on/off) at the beginning of subproblem k, and on the other hand, the expected component's profit if operated. We estimate the profit of component s for each time step of the full time horizon. The profit calculation is set up once for each component in the system. Afterwards, the profit calculation can be used for each subsequent operational optimization (only the energy prices have to be updated).

We define the profit of the component s as the difference between its operation cost and the operation cost of every other component s' that can fulfill the same demand.

$$profit^{\max/\min}_{s,s',t} = c_{s'}, t \cdot I^{\max/\min}_{s,s'} - c_{s,t} \cdot I^{\max/\min}_{s}, \quad \forall t \in \mathrm{T}^{\text{full}}, s \in S, s' \in S^{\max/\min}_{s}. \tag{5.16}$$

$c_{s,t}$ is the specific price for the input energy of component s in time step t, e.g., gas or electricity price. The specific price $c_{s,t}$ has to be known for all components that cause start-up costs or can replace components that cause start-up costs. Since the components' operation is unknown before operational optimization, we do not know any component's load. Thus, we calculate the profit for both the maximum load and the minimum load of component s. $I^{\max/\min}_{s}$ is the input energy of component s at either maximum or minimum load. $I^{\max/\min}_{s,s'}$ is the input energy of component s' necessary to provide the same output as component s.

We sum the profit of the component s compared to s' (Equation (5.16)) over time to get the cumulative profit $cp_{s,s',t}$ for the specific pair of s and s' (Figure 5.3):

$$cp_{s,s',t} = \sum_{\tilde{t}=0}^{t} profit^{\max/\min}_{s,s',\tilde{t}}. \tag{5.17}$$

The set CP_s includes the cumulative profits of the component s compared to all its alternative components. The Adaptive Rolling Horizon chooses the maximum of all foresights for component s in subproblem k as the necessary foresight:

$$\mathrm{T}^{\text{fore}}_{k,s} = \max_{CP_s} \mathrm{T}^{\text{fore}}_{k,s,cp}. \tag{5.18}$$

For better readability, we neglect the indices of $cp_{s,s',t}$ and use cp instead for a specific cumulative profit from the set of all possible cumulative profits CP_s in the remainder of this paper.

In the following, we explain how Adaptive Rolling Horizon determines the foresight $\mathrm{T}^{\text{fore}}_{k,s,cp}$ for any cumulative profit. The goal is to ensure the correct start-ups of the

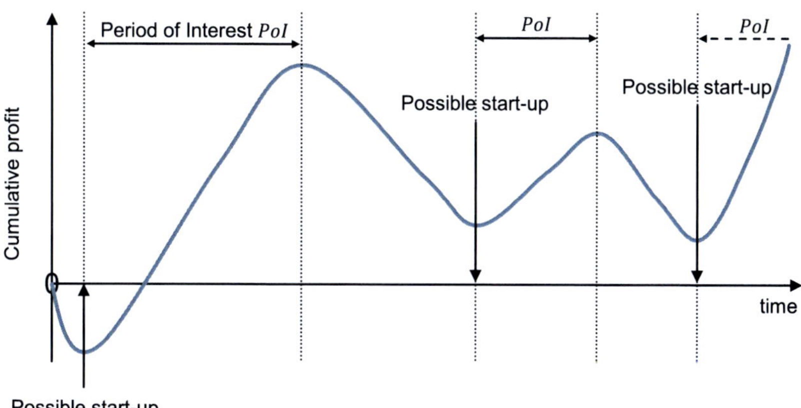

Figure 5.3: Cumulative profit of a generic component s compared to alternative component s' for the full time horizon of an operational optimization. The cumulative profit is the integral of the profit of the component s compared to s' (Equation (5.17)). The periods of rising cumulative profits are called Periods of Interest PoI. At the beginning of each Period of Interest PoI, starting up the component s' can be economically beneficial.

components in the rolling horizon, i.e., the same start-up as in the full time horizon. Thereby, we also ensure the correct shut-downs because a correct start-up of a component s simultaneously marks a correct shut-down of the alternative component s'. As mentioned above, ensuring correct start-ups is a strong requirement for the method and leads to a complex case distinction.

Figure 5.3 shows the cumulative profit of a component s compared to its alternative component s'. Operating component s is cheaper than operating the component s' during the time of rising cumulative profit indicated by a positive slope. We call the periods of rising cumulative profit the Periods of Interest PoI. At the beginning of each Period of Interest PoI, starting up the component s can be economically beneficial. In reality, the components may start up and shut down at any time due to, e.g., changes in demands, minimum up- and downtimes, etc. The Adaptive Rolling Horizon accounts for start-ups and shut-downs at any time. However, here, we simplify the description of the Adaptive Rolling Horizon by accounting only for start-ups at the beginning of a Period of Interest PoI or the beginning of a subproblem k. Each minimum of the cumulative profit marks the beginning of a Period of Interest PoI and, thus, a possible start-up.

Before calculating the foresight $T^{\text{fore}}_{k,s,cp}$ for any cumulative profit cp, the Adaptive Rolling Horizon checks if starting up the component s can be excluded in subproblem k due to constraints such as minimum up- and downtimes. If starting up the component s cannot be excluded beforehand, the Adaptive Rolling Horizon applies the algorithm to calculate the foresight $T^{\text{fore}}_{k,s,cp}$ (Figure 5.4). The algorithm first identifies if starting up the component s is economically beneficial or not, and subsequently determines a suitable foresight for the identified case.

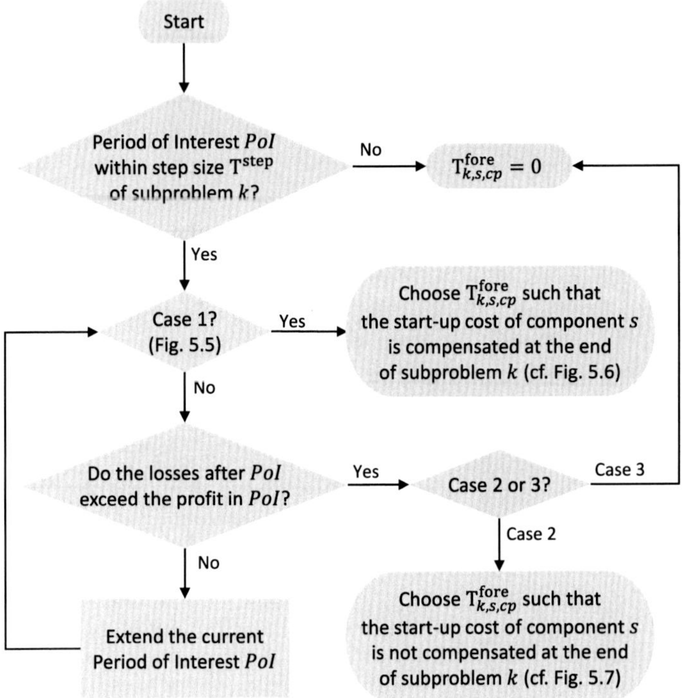

Figure 5.4: Scheme of the algorithm to calculate the foresight $T^{\text{fore}}_{k,s,cp}$ for the cumulative profit cp of the component s in subproblem k.

At first, the algorithm checks if the step size T^{step} of subproblem k overlaps with a Period of Interest PoI. If not, starting up the component s is not economically beneficial. In this case, we set the foresight $T^{\text{fore}}_{k,s,cp} = 0$.

Suppose the step size $\mathrm{T}^{\mathrm{step}}$ contains a Period of Interest PoI. In that case, we analyze the time from the beginning of this Period of Interest PoI until the beginning of the next Period of Interest PoI and check if one of the following three cases occurs (Figure 5.5):

1. Case 1: The Period of Interest PoI generates sufficient profit to compensate the sum of the start-up cost of the component c_s^{start} and the start-up cost of the alternative component $c_{s'}^{\mathrm{start}}$. In this case, starting up component s is economically beneficial.
2. Case 2: The Period of Interest PoI generates sufficient profit to compensate the start-up cost of the component c_s^{start}. However, the following losses exceed the profit from the Period of Interest PoI. In this case, starting up component s is not economically beneficial.
3. Case 3: The Period of Interest PoI does not generate sufficient profit to compensate the start-up cost c_s^{start} of component s. The following losses exceed the profit from the Period of Interest PoI. In this case, starting up component s is not economically beneficial.

This distinction is necessary to determine a suitable foresight $\mathrm{T}_{k,s,cp}^{\mathrm{fore}}$ as we show in the following. Starting up the component s is only guaranteed to be economically beneficial if we compensate the start-up costs of components s and s' (case 1, Figure 5.5). If we started up component s, compensated only its start-up cost, and started up component s' after the current Period of Interest PoI, we would have to pay the start-up costs of both components. Still, we only compensate the start-up cost of component s. In this case, it would have been more economical not to start up component s at the beginning of the current Period of Interest PoI.

However, if the current Period of Interest PoI compensates the start-up costs of the components s and s' (case 1), starting up the component s at the beginning of the current Period of Interest PoI is preferable independent from what happens later. In case 1, we set the foresight $\mathrm{T}_{k,s,cp}^{\mathrm{fore}}$ such that the start-up of component s pays off in subproblem k. We calculate the time step $t_{k,s,cp}^{\mathrm{BE}}$ ("break-even" point) at which the start-up of component s pays off in subproblem k (Figure 5.6). The foresight $\mathrm{T}_{k,s,cp}^{\mathrm{fore}}$ is chosen such that subproblem k contains time step t_k^{BE}, and is thus given by

$$\mathrm{T}_{k,s,cp}^{\mathrm{fore}} = \max(t_{k,s,cp}^{\mathrm{BE}} - t_k^{\mathrm{end,step}}, 0), \tag{5.19}$$

with the time step $t_{k,s,cp}^{\mathrm{BE}}$ defined by

$$\sum_{t^{\mathrm{start,PoI}}}^{t_{k,s,cp}^{\mathrm{BE}}} profit_{s,s',t} > c_s^{\mathrm{start}} \,, \quad \sum_{t^{\mathrm{start,PoI}}}^{t_{k,s,cp}^{\mathrm{BE}}-1} profit_{s,s',t} \leq c_s^{\mathrm{start}}. \tag{5.20}$$

$t^{\mathrm{start,PoI}}$ is the first time step of the current Period of Interest PoI. $t_k^{\mathrm{end,step}}$ is the last time step of the step size $\mathrm{T}^{\mathrm{step}}$ in subproblem k.

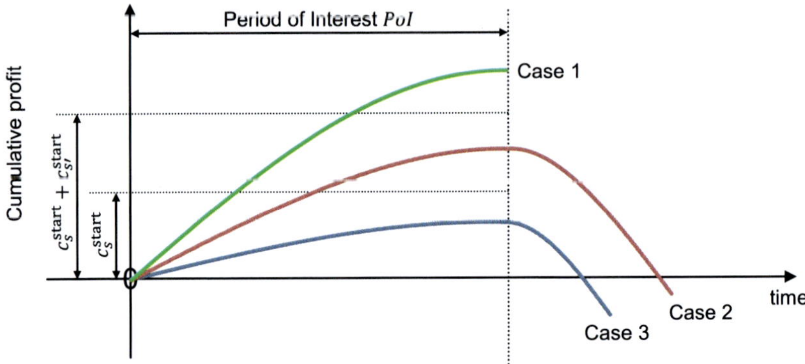

Figure 5.5: The three cases that can occur if the step size $\mathrm{T}^{\mathrm{step}}$ contains a Period of Interest PoI. In case 1 (green), starting up the component s is economically beneficial. In cases 2 (red) and 3 (blue), starting up the component s is not economically beneficial.

If case 1 does not occur, we check for cases 2 or 3. For this purpose, we analyze the losses following the current Period of Interest PoI. Case 2 or 3 occurs if the losses exceed the profit from the Period of Interest PoI, and case 1 is already excluded (Figure 5.7). In both cases, starting up the component s at the beginning of the current Period of Interest PoI is not economically beneficial. Thus, for cases 2 and 3, we determine the foresight $\mathrm{T}_{k,s,cp}^{\mathrm{fore}}$ such that component s is not wrongly started in the current Period of Interest PoI.

In case 3, the Period of Interest PoI does not compensate the start-up cost for the component s at any time. Thus, we can set the foresight $\mathrm{T}_{k,s,cp}^{\mathrm{fore}} = 0$. However, in case 2, the Period of Interest PoI compensates the start-up cost for the component s, although starting up the component s is not economically beneficial. Thus, we set the foresight $\mathrm{T}_{k,s,cp}^{\mathrm{fore}}$ such that the start-up of component s does not pay off in

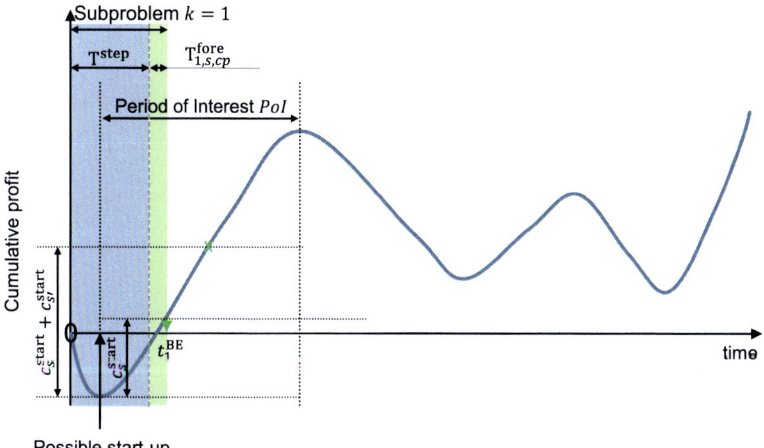

Figure 5.6: The determination of the foresight $T^{fore}_{k,s,cp}$ is shown for subproblem $k = 1$. The Period of Interest PoI occurs within the step size T^{step} of subproblem 1. The Period of Interest PoI compensates the start-up costs of the components s and s'. Thus, starting up the component s at the beginning of the Period of Interest PoI is economically beneficial. The foresight $T^{fore}_{1,s,cp}$ is chosen such that t_1^{BE} is the last time step in subproblem $k = 1$.

the current subproblem k (Figure 5.7). A further requirement is that the foresight $T^{fore}_{k,s,cp}$ exceeds the current Period of Interest PoI. Otherwise, the foresight of another component could lead to a start-up of the component s paying off in subproblem k (Equation (5.15)). The foresight is then given by

$$T^{fore}_{k,s,cp} = \max(t^{BE}_{k,s,cp} - t_k^{end,step}, 0), \tag{5.21}$$

with the time step $t^{BE}_{k,s,cp}$ defined by

$$\sum_{t^{start,PoI}}^{t^{BE}_{k,s,cp}} profit_{s,s',t} < c_s^{start}\,, \quad \sum_{t^{start,PoI}}^{t^{BE}_{k,s,cp}-1} profit_{s,s',t} \geq c_s^{start}. \tag{5.22}$$

If none of the three cases occurs, the losses after the current Period of Interest PoI do not exceed the profit (Figure 5.8). In this case, we cannot determine a suitable foresight $T^{fore}_{k,s,cp}$ based only on information about the current Period of Interest PoI.

Starting up the component s can still be economical if the start-up pays off later. Thus, we extend the current Period of Interest PoI to the end of the next Period of Interest PoI (Figure 5.8). For this extended Period of Interest PoI, we analyze the three cases again (Figure 5.5). Figure 5.8 illustrates case 1 for an extended Period of Interest PoI. This extension of the Period of Interest PoI continues until either one of the three cases occurs or the extended Period of Interest PoI reaches the end of the full time horizon. In this way, the foresight $\mathrm{T}^{\text{fore}}_{k,s,cp}$ is determined for every alternative component s'. Furthermore, the user can specify a maximum extension of Period of Interest PoI to limit the foresight $\mathrm{T}^{\text{fore}}_{k,s,cp}$. The foresight $\mathrm{T}^{\text{fore}}_{k,s,cp}$ is set to 0 if the maximum extension is reached and none of the three cases occurs. If the user specifies a maximum extension, wrong start-ups and shut-downs can occur.

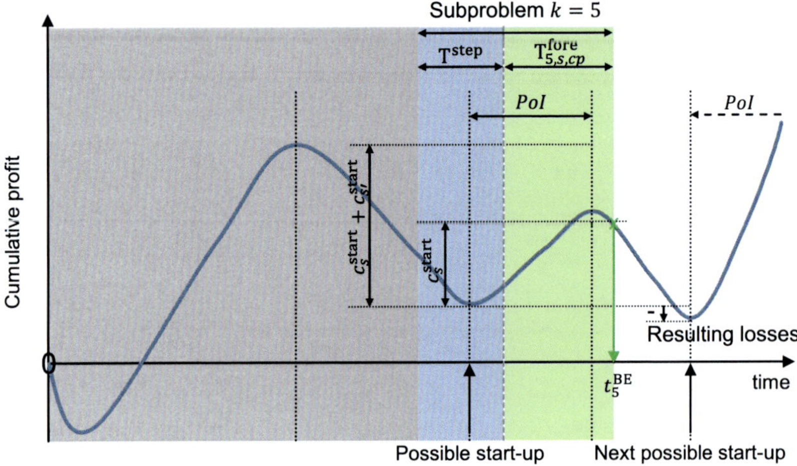

Figure 5.7: The determination of the foresight $\mathrm{T}^{\text{fore}}_{k,s,cp}$ is shown for subproblem $k = 5$. The Period of Interest PoI starts within the step size T^{step} of subproblem 5. The Period of Interest PoI compensates the start-up costs of the component s. However, starting up the component s at the beginning of the Period of Interest PoI is not economically beneficial because of the resulting losses. Thus, the foresight $\mathrm{T}^{\text{fore}}_{5,s,cp}$ is chosen such that starting up the component s does not pay off in subproblem 5. t_5^{BE} is the last time step in subproblem $k = 5$.

The maximum value over all components is finally chosen as foresight T_k^{fore} (Equation (5.15)). This foresight T_k^{fore} is as short as possible but as long as necessary to decide correctly about start-ups and shut-downs for all components during optimization of subproblem k.

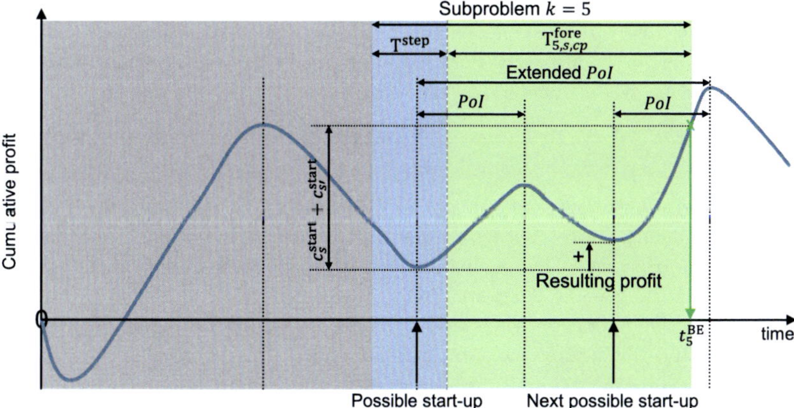

Figure 5.8: The determination of the foresight $T_{k,s,cp}^{\mathrm{fore}}$ is shown for subproblem $k = 5$ (cf. Figure 5.7). The Period of Interest PoI starts within the step size T^{step} of subproblem 5. For the current Period of Interest PoI, none of the three cases occurs. Thus, we analyze the extended Period of Interest PoI. For the extended Period of Interest PoI, case 1 occurs, and thus the foresight $T_{5,s,cp}^{\mathrm{fore}}$ is chosen such that starting up the component s pays off in subproblem 5. t_5^{BE} is the last time step in subproblem $k = 5$.

5.3 Case Study: real-world multi-energy system

The Adaptive-Rolling-Horizon approach is applied to the operational optimization of a real-world multi-energy system. The multi-energy system supplies steam demands on two pressure levels, a cooling demand, an electricity demand, and a compressed-air demand. The multi-energy system consists of multiple coal-driven and gas-driven boilers, a gas turbine combined with a heat recovery boiler, and multiple steam turbines.

We solve one operational optimization problem for each season (winter, spring, summer, autumn). Each operational optimization problem has a full time horizon of four weeks with hourly resolution. The operational optimizations minimize the operational expenditure $OPEX$. We solved all problems using Gurobi 9.0.0 (Gurobi Optimization, 2020) in Python with the standard relative MIP gap of 10^{-4} and a time limit of 10 h.

We compare the computation time and the solution quality of the Adaptive-Rolling-Horizon approach with the common Rolling Horizon and Perfect Foresight. Perfect Foresight refers to solving the original optimization problem. For the common Rolling Horizon, we use two parameter sets for the step size $\mathrm{T}^{\mathrm{step}}$ and the foresight $\mathrm{T}^{\mathrm{fore}}$:

1. Step size $\mathrm{T}^{\mathrm{step}} = 5$ time steps and foresight $\mathrm{T}^{\mathrm{fore}} = 10$ time steps: this parameter set has been found heuristically in practice to reach a short computation time.
2. Step size $\mathrm{T}^{\mathrm{step}} = 20$ and foresight $\mathrm{T}^{\mathrm{fore}} = \max_k \mathrm{T}^{\mathrm{fore}}_k$: This parameter set refers to a Rolling-Horizon optimization with the maximum foresight from Adaptive Rolling Horizon. This parameter set is only known if all necessary foresight calculations for the Adaptive Rolling Horizon have already been performed. Thus, this parameter set is only used to benchmark the performance of the Adaptive-Rolling-Horizon optimization.

First, to show the advantage of the Adaptive Rolling Horizon, we analyze the cumulative solver times of the Adaptive Rolling Horizon and the Rolling Horizon with the parameter set $\mathrm{T}^{\mathrm{step}} = 20$, $\mathrm{T}^{\mathrm{fore}} = \max_k \mathrm{T}^{\mathrm{fore}}_k$ (spring (top) and summer (bottom), Figure 5.9). The Adaptive Rolling Horizon determines a maximum foresight $\mathrm{T}^{\mathrm{fore}}_{22} = 35$ time steps in spring and a maximum foresight $\mathrm{T}^{\mathrm{fore}}_{1} = 90$ time steps in summer. Thus, in the common Rolling Horizon, we set $\mathrm{T}^{\mathrm{fore}} = 35$ time steps for all subproblems in spring and $\mathrm{T}^{\mathrm{fore}} = 90$ time steps in summer. The Adaptive Rolling Horizon solves all subproblems faster or at least as fast as the common Rolling Horizon. The total computation time of the Adaptive Rolling Horizon is nearly two times faster in spring (144 s vs. 271 s) and nearly three times faster in summer (183 s vs. 493 s).

Furthermore, the average foresight $\overline{\mathrm{T}^{\mathrm{fore}}}$ of the Adaptive Rolling Horizon changes with the season. In spring, the average foresight $\overline{\mathrm{T}^{\mathrm{fore}}}$ contains 27 time steps, whereas it contains 46 time steps in summer. This change in the average foresight $\overline{\mathrm{T}^{\mathrm{fore}}}$ suggests that a fixed foresight $\mathrm{T}^{\mathrm{fore}}$ is not suitable.

Instead, adapting the foresight for each subproblem k reduces computation times while retaining high solution quality. For the Adaptive Rolling Horizon, the cumulative computation time to calculate all foresights for all subproblems is only 1 s per operational optimization, i.e., 0.03 s per subproblem. In all foresight calculations for the case study, the Period of Interest PoI is extended once per subproblem at maximum.

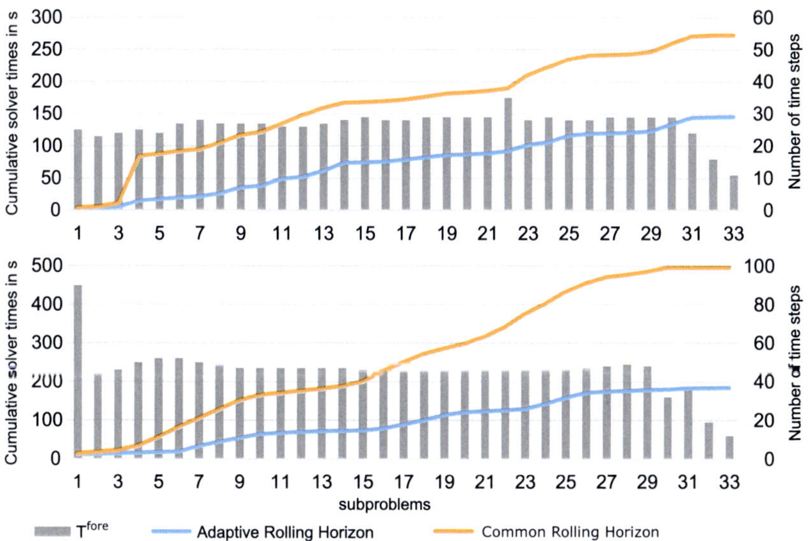

Figure 5.9: Cumulative solver times for the Adaptive Rolling Horizon and the common Rolling Horizon with the parameter set $\mathrm{T}^{\mathrm{step}} = 20, \mathrm{T}^{\mathrm{fore}} = \max_k \mathrm{T}_k^{\mathrm{fore}}$ for a full time horizon of four weeks in spring (top) and summer (bottom). Additionally, the number of time steps in the foresight $\mathrm{T}_k^{\mathrm{fore}}$ of the Adaptive Rolling Horizon is shown for each subproblem. The Adaptive Rolling Horizon solves all subproblems faster and, thus, provides a solution for the four weeks nearly two times faster in spring and nearly three times faster in summer.

Figure 5.10 shows the relative increase in $OPEX$ related to the $OPEX$ of the Perfect-Foresight solutions over the total computation time for all four seasons and the four solution approaches Adaptive Rolling Horizon, common Rolling Horizon with both parameter sets, and Perfect Foresight. The Adaptive Rolling Horizon reaches excellent solution quality in short computation times. Compared to Perfect Foresight, the increase in $OPEX$ is less than 0.1 % on average, whereas the computation time is more than 150 times shorter on average.

The common Rolling Horizon with the parameter set $\mathrm{T}^{\mathrm{step}} = 20$ and $\mathrm{T}^{\mathrm{fore}} = \max_k \mathrm{T}_k^{\mathrm{fore}}$ reaches the same solution quality as the Adaptive Rolling Horizon (increase in $OPEX$ is less than 0.1 % on average compared to Perfect Foresight). However, it

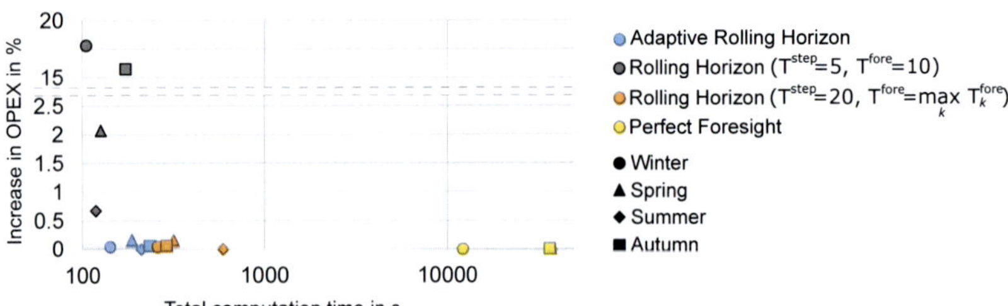

Figure 5.10: Total computation times and relative increase in $OPEX$ for all approaches and seasons. For each season, the increase in $OPEX$ is related to the $OPEX$ of the Perfect-Foresight solution.

provides the solution nearly two times slower than the Adaptive Rolling Horizon. The common Rolling Horizon with the parameter set $\mathrm{T}^{\mathrm{step}} = 5$ and $\mathrm{T}^{\mathrm{fore}} = 10$ (grey) reaches the shortest computation time of all approaches (23 % faster than Adaptive Rolling Horizon on average). However, the solution quality is the worst among all considered approaches. $OPEX$ increases by more than 9 % on average compared to Perfect Foresight.

Obviously, Perfect Foresight (yellow) reaches the best solution quality. However, the total computation time is the highest. The time limit of 10 h is reached in three of the four seasons and is therefore not always satisfying for practical applications.

In summary, the Adaptive Rolling Horizon decreases the total computation time by decomposing the original optimization problem and adapting the foresight for each subproblem. Still, the Adaptive Rolling Horizon retains excellent solution quality for the operational optimization problems.

5.4 Conclusion

The Adaptive-Rolling-Horizon approach is proposed to decrease the computation time of operational optimization while ensuring excellent solution quality. The approach adaptively determines an individual foresight for each subproblem based on knowledge about the system and time-dependent parameters (e.g., time series of energy prices).

We detail the approach for a multi-energy system to exemplify how a suitable foresight $\mathrm{T}_k^{\mathrm{fore}}$ can be identified. Identifying a suitable foresight uses knowledge about the specific system. Thus, the determination of the foresight $\mathrm{T}_k^{\mathrm{fore}}$ probably needs to be adapted for other problems than short-term operational optimization of multi-energy systems.

In the case of the multi-energy system, the foresight is determined for each subproblem by considering each component's profit and start-up cost. With this information, the Adaptive Rolling Horizon estimates the necessary foresight for each subproblem to decide correctly about start-ups and shut-downs. However, the considered multi-energy system does not include energy storage units. Thus, investigating the influence of energy storage units on a suitable foresight $\mathrm{T}_k^{\mathrm{fore}}$ is a possible way to continue this work.

In a real-world case study, the Adaptive Rolling Horizon is not able to solve operational optimizations in a reliably short time as the computation times between specific instances differ significantly. However, the Adaptive Rolling Horizon substantially reduces the computation time by a factor of 150 on average compared to Perfect Foresight. Still, the approach provides solutions near the globally optimal solution, increasing $OPEX$ by less than 0.1 % on average. The results encourage adjusting the foresight in a Rolling-Horizon optimization adaptively by employing knowledge of the system to be optimized.

Chapter 6

Boosting Operational Optimization by Artificial Neural Nets

In this chapter, we propose a decomposition method to solve the operational optimization of multi-energy systems using artificial neural nets. Similar to the Adaptive Rolling Horizon from the previous chapter, the proposed method copes with the second major task identified in Chapter 2.4: finding high-quality solutions for operational optimization of multi-energy systems in short times. In contrast to the Adaptive Rolling Horizon, no expert knowledge about the multi-energy system is necessary to employ this method. Furthermore, the decomposition method can be applied to multi-energy systems containing energy storage units. We show that the proposed method is able to solve the operational optimization of multi-energy systems in a reliably short time. Due to this reliably short time, the decomposition method enables compliance with strict time limits for repeated operational optimization.

In Section 6.1, we extend the MILP operational optimization problem of multi-energy systems by constraints for energy storage units. In Section 6.2, we present the decomposition method for boosting the operational optimization by neural nets. In Section 6.3, we apply the decomposition method to a real-world case study and a literature case study. In Section 6.4, we conclude with the key findings.

Major parts of this chapter are reproduced from:

Kämper, A., Delorme, R., Leenders, L., and Bardow, A. (2023). Boosting Operational Optimization of Multi-Energy Systems by Artificial Neural Nets. *Computers & Chemical Engineering*, 173, 108208.

Contribution report: Principal author, Conceptualization, Methodology, Software, Validation, Formal Analysis, Investigation, Data Curation, Writing - Original Draft, Visualization, Project administration.

6.1 Extension of the MILP operational optimizaton problem

The proposed decomposition method can be applied to multi-energy systems containing energy storage units. Thus, we extend the MILP operational optimization problem of multi-energy systems from Chapter 5.1 by constraints for energy storage units. These constraints are given by:

$$L_{st,e,t} \quad L_{st,e,t-1} \quad -I_{st,e,t}\eta_{st}^{\text{load}}\Delta \mathrm{t}_t - \frac{O_{st,e,t}}{\eta_{st}^{\text{unload}}}\Delta \mathrm{t}_t, \quad \forall st \in ST, e \in E_{st}, t \in T \setminus \{0\}, \tag{6.1}$$

$$L_{st,e,t} \leq l_{st,e}^{\max}, \quad \forall st \in ST, e \in E_{st}, t \in T, \tag{6.2}$$

$$L_{st,e,t} \geq l_{st,e}^{\min}, \quad \forall st \in ST, e \in E_{st}, t \in T, \tag{6.3}$$

$$I_{st,e,t} < i_{st}^{\max}\delta_{st,t}^{\text{load}}, \quad \forall st \in ST, e \in E_{st}, t \in T, \tag{6.4}$$

$$O_{st,e,t} \leq o_{st}^{\max}\delta_{st,t}^{\text{unload}}, \quad \forall st \in ST, e \in E_{st}, t \in T, \tag{6.5}$$

$$1 \geq \delta_{st,t}^{\text{load}} + \delta_{st,t}^{\text{unload}}, \quad \forall st \in ST, t \in T, \tag{6.6}$$

Equations (6.1) describe the storage level $L_{st,e,t}$ for all storage units st and energy forms e at time step t. E_{st} is the set of energy forms stored by storage unit st. Equations (6.2) and (6.3) specify the maximum and minimum storage level $l_{st,e}^{\max}$ and $l_{st,e}^{\min}$. Equations (6.4) and (6.5) restrict the input and output energy flows of storage st to a maximum input flow $i_{st}^{\max}$ and a maximum output flow $o_{st}^{\max}$. Equations (6.6) ensure that storage st cannot be loaded and unloaded simultaneously.

The decomposition method proposed in the following solves operational optimization problems consisting of the objective function in Equation (5.1) and the constraints in Equations (5.2) to (5.14) and (6.1) to (6.6).

6.2 Decomposition method for boosting operational optimization

The proposed decomposition method aims at solving the extended operational optimization problem (Section 6.1) in a reliably short time while providing a feasible and sufficiently accurate solution. The reliably short computation time is reached by decomposing the original problem into single-time-step optimization problems. The main drawback of single-time-step optimizations is that they do not consider any future information about demands or prices, which can lead to poor solution qualities and infeasibilities. To tackle this drawback, the decomposition method uses artificial neural nets (ANNs). The ANNs consider past, current, and future demands and prices to predict binary and continuous decision variables of the next single-time-step optimization. The ANNs are trained on long-term MILP solutions. Thereby, the ANNs can incorporate long-term decisions into the operational optimization of a single time step. We summarize the predicted binary decision variables in a vector of binary variables $\boldsymbol{\delta}$ and the predicted continuous decision variables in a vector of continuous variables $\boldsymbol{x}$. The decomposition method queries the ANNs once for each time step to be optimized and employs the predictions of the ANNs in the corresponding single-time-step optimization. The premise of the proposed decomposition method is that the original MILP problem (Section 6.1) can be solved. If the original MILP cannot be solved, we cannot generate the necessary training data such that the decomposition method cannot be employed.

In Section 6.2.1, we present the general workflow of the decomposition method. In Section 6.2.2, we describe how to acquire the necessary data for the ANN training. In Section 6.2.3, we explain the architectures, hyperparameter tuning, and training of the ANNs.

6.2.1 General workflow of the decomposition method

The time period of the operational optimization solved by the decomposition method can be chosen arbitrarily. In the following, we describe the decomposition method for solving the operational optimization of one week with an hourly resolution (168 time steps), minimizing the operational expenditures $OPEX$ (cf. Equation 5.1). The first time step of the operational optimization is denoted by t^{start}, and the last by t^{end}.

In general, multi-energy systems already operate before an operational optimization. These operating states are considered as initial conditions at the beginning of an operational optimization. We summarize the operating states of all system components under the term *system operation* in the following.

The workflow of the decomposition method is shown in Figure 6.1. As input, the method requires the demands, prices, and system operation of the past week as well as the forecasts of demands and prices for the current week. With the given information, two separate ANNs predict decision variables for time step t. ANN^{class} performs a classification task and predicts the binary decision variables $\boldsymbol{\delta}_t^{\text{ANN}}$, whereas ANN^{approx} performs an approximation task and predicts the continuous decision variables $\boldsymbol{x}_t^{\text{ANN}}$.

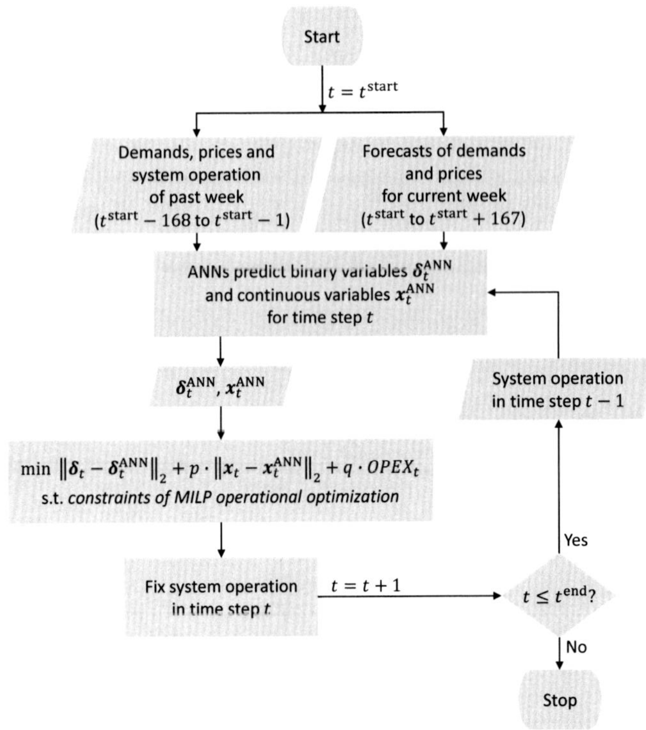

Figure 6.1: Flowchart of the proposed decomposition method.

However, the ANNs do not guarantee that the MILP operational optimization problem of time step t is feasible with fixed predicted decision variables $\boldsymbol{\delta}_t^{\text{ANN}}$ and $\boldsymbol{x}_t^{\text{ANN}}$. To ensure that the ANN predictions lead to a feasible solution, the method solves a mixed-integer quadratic programming (MIQP) problem for time step t. In the MIQP, the ANN predictions are parameters. The objective function of the MIQP considers the ANN predictions hierarchically because correctly predicting the on/off states of the components is more important than precisely predicting the loads of the components:

$$\min \left(\left|\left|\boldsymbol{\delta}_t - \boldsymbol{\delta}_t^{\text{ANN}}\right|\right|_2 + p \cdot \left|\left|\boldsymbol{x}_t - \boldsymbol{x}_t^{\text{ANN}}\right|\right|_2 + q \cdot OPEX_t \right) \quad . \tag{6.7}$$

The objective function (6.7) ensures that the binary decision variables $\boldsymbol{\delta}_t$ and the continuous decision variables $\boldsymbol{x}_t$ are as close as possible to those predicted by the ANNs $\boldsymbol{\delta}_t^{\text{ANN}}$ and $\boldsymbol{x}_t^{\text{ANN}}$. The continuous variables $\boldsymbol{x}_t$ and $\boldsymbol{x}_t^{\text{ANN}}$ are weighted by the parameter p such that their weight in the objective function is smaller than the weight of the binary variables by three orders of magnitude. Thus, the value of the parameter p depends on the order of magnitude of the continuous variables $\boldsymbol{x}_t$. $OPEX_t$ represents the objective of the original MILP operational optimization (Equation 5.1) in time step t:

$$\min OPEX = \sum_{t \in T} (OPEX_t) \,. \tag{6.8}$$

The presence of the term $OPEX_t$ in the objective function of the MIQP aims to choose optimal values for variables that the ANNs do not predict rather than feasible but otherwise randomly chosen values. However, $OPEX_t$ shall not influence the choice of the values of $\boldsymbol{x}_t$ and $\boldsymbol{\delta}_t$. For this purpose, its weighting parameter q is chosen such that

$$q \cdot OPEX_t \ll \left|\left|\boldsymbol{\delta}_t - \boldsymbol{\delta}_t^{\text{ANN}}\right|\right|_2 + p \cdot \left|\left|\boldsymbol{x}_t - \boldsymbol{x}_t^{\text{ANN}}\right|\right|_2 . \tag{6.9}$$

Thereby, the influence of $OPEX_t$ on the values of $\boldsymbol{x}_t$ and $\boldsymbol{\delta}_t$ is negligible. Thus, the MIQP yields a feasible solution for time step t while keeping the binary and continuous decision variables $\boldsymbol{\delta}_t$ and $\boldsymbol{x}_t$ close to the predictions of the ANNs. The constraints of the MIQP are the same as in the MILP operational optimization of time step t (Equations (5.2) to (5.14)). Since the ANNs predict the decision variables based on long-term MILP optimizations, the method considers long-term effects in an MIQP of a single time step. Thereby, the decomposition method resolves the described main drawback of single-time-step optimizations.

The solution for time step t is saved and used as input for the ANNs to predict the decision variables $\boldsymbol{\delta}_{t+1}$ and $\boldsymbol{x}_{t+1}$ for time step $t+1$ (Figure 6.1). Hence, the ANNs and MIQP work together within an iterative solution method to derive a solution for the time period from t^{start} to t^{end}.

However, the decomposition method cannot guarantee a feasible solution for the operational optimization from t^{start} to t^{end} if time-coupling constraints are present like minimum up- and downtimes, ramping constraints, and minimum part loads. These constraints strongly restrict the flexibility of the system such that the system cannot react to high demand variations from one time step to the next. Still, the decomposition method is able to provide high-quality solutions even for a strongly restricted multi-energy system in most cases, as we show in Section 6.3.

Furthermore, the high precision of the ANN predictions for binary variables allows for an optional extension of the decomposition method. While the solution of the decomposition method can be used directly, it can also be used to reduce the computational effort of the original optimization problem. The effort is reduced by fixing the values of the predicted binary variables to the values provided by the decomposition method. The extension improves the quality of the solution as demonstrated in Section 6.3.3.

6.2.2 Data acquisition for training the artificial neural nets

To apply the proposed decomposition method, we have to train the ANNs. Training ANNs requires large data sets of input and target pairs. Here, the ANNs train on the solutions provided by MILP optimizations of the given multi-energy system. As mentioned before, we illustrate the data acquisition for solving the operational optimization of one week with hourly resolution. The inputs of the trained ANNs are the time series of the system operation, energy demands, and prices from the past week, as well as the forecasts for energy demands and prices for current week, including the current time step. In the decomposition method, the ANNs predict the binary and continuous decision variables $\boldsymbol{\delta}_t^{\mathrm{ANN}}$ and $\boldsymbol{x}_t^{\mathrm{ANN}}$ for each time step of the current week. Thus, we solve MILP problems for two consecutive weeks to generate the input – target pairs for ANN training. Here, we choose two consecutive weeks such that the decomposition method has one week of information from the past for each time step t to be predicted. Thereby, the ANNs can consider long-term effects of up to one week, e.g., component up- and downtimes of multiple days. The MILP optimizations can be performed offline and in parallel, allowing less restrictive time limits and a repeated update of the ANNs.

A potential shortcoming is that the training requires historical time series. If the available historical time-series data is insufficient, we propose using a vector autoregressive model to generate synthetic time series for demands and prices from available historical time series (Klöckl and Papaefthymiou, 2010). Generating synthetic time series by the method of Klöckl and Papaefthymiou (2010) has been shown to provide a sufficiently large number of samples to evaluate a wide range of possible system operations (Efstratiadis et al., 2014).

6.2.3 Architecture, hyperparameter tuning, and training of the artificial neural nets

Before describing the ANN training, we discuss the ANN architecture and hyperparameter tuning. ANNs' architectures always have to be tailored to their specific application (Goodfellow et al., 2016). Thus, designing ANNs for new applications is not straightforward.

For the proposed decomposition method, we design two artificial neural nets: $\text{ANN}^{\text{approx}}$ and $\text{ANN}^{\text{class}}$. $\text{ANN}^{\text{approx}}$ predicts continuous values for the subsequent MIQP problem, which solves the operational optimization of a multi-energy system for time step t. Thus, $\text{ANN}^{\text{approx}}$ solves an approximation problem. $\text{ANN}^{\text{class}}$ predicts binary values for the subsequent MIQP problem. The binary values describe the on/off states of components in the multi-energy system. The on/off states of the components are independent from each other. Thus, the $\text{ANN}^{\text{class}}$ solves a multi-label classification problem (Read et al., 2009).

Although the ANNs solve different problems, their inputs are the same time series. For this reason, we choose the same *hyper architecture* for both ANNs. The hyper architecture represents the set of all possible ANN architectures and is described in the following (Figure 6.2). We split the hyper architecture of the ANNs into three sections: past, present (time step t), and future. In the section that corresponds to the past time before time step t, one input layer takes the time series of demand and prices, and a second input layer takes the time series of system operation from the past week ($t - 168$ to $t - 1$). The input layers pass the input time series into a concatenating layer. The concatenating layer concatenates the input time series to a single tensor.

In the section that corresponds to the future time after time step t, the time series of the demand and price forecasts for the next week ($t + 1$ to $t + 167$) are processed. However, as the decomposition method progresses, the time series of the demand and price forecasts after time step t shortens. For this reason, we extend the time series of the demand and price forecasts with masking values such that the input layer always receives time series with 167 values. The following masking layer masks out all time steps with masking values as they contain no information about the future. For the sections reflecting past and future, we include LSTM (long short-term memory) layers since they can learn the structure of time series (Hochreiter and Schmidhuber, 1997). Specifically, we use so-called *many-to-one* LSTMs to encode the input time series of past and future in one single state, respectively. The LSTM layers are followed by dense feedforward layers. We use dense feedforward layers with ReLU (Rectifier

Linear Unit) activation functions (Glorot et al., 2011) to increase the capacity of the ANNs while retaining a good training performance (LeCun et al., 2015). The input at time step t is directly fed into dense feedforward layers since it is not a time series and no compression by an LSTM is needed. The reshape layers in the past and future sections enable the subsequent concatenation of the information from all three sections. Afterwards, the concatenated information enters the final dense feedforward layers. The last dense feedforward layer consists of as many neurons as the number of predicted decision variables. The output of that layer is the output of the ANN.

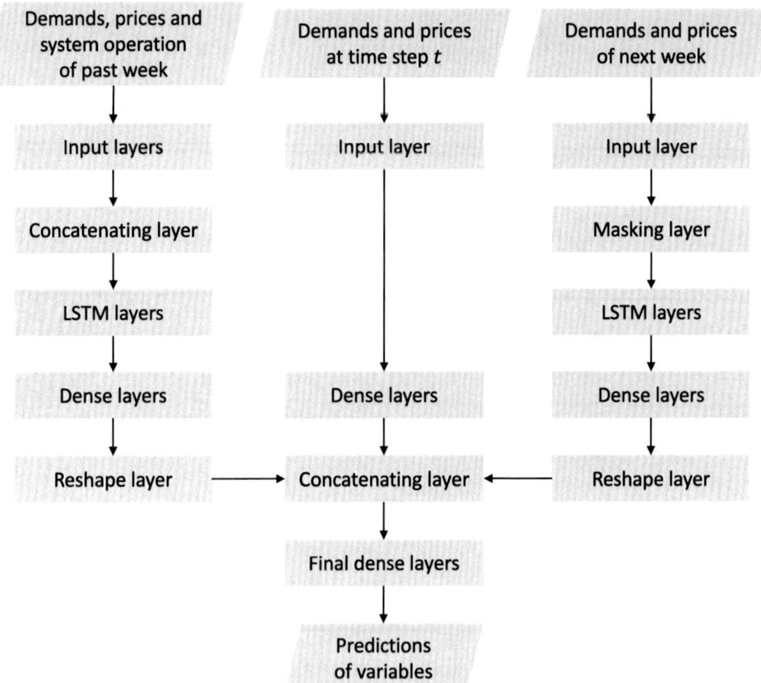

Figure 6.2: Hyper architecture of the ANNs. The inputs for the past and future (left and right parts) are time series. For this reason, these parts include LSTM (long short-term memory) layers. LSTM blocks can learn the structure of time series. The ANNs are implemented in TensorFlow.

Based on the described hyper architecture, we perform a hyperparameter tuning using Keras tuner 1.0.1 (O'Malley et al., 2019) separately for both ANNs to find suitable ANN architectures for the application in this work (Figure 6.3). A specific set of hyperparameters defines a specific ANN architecture.

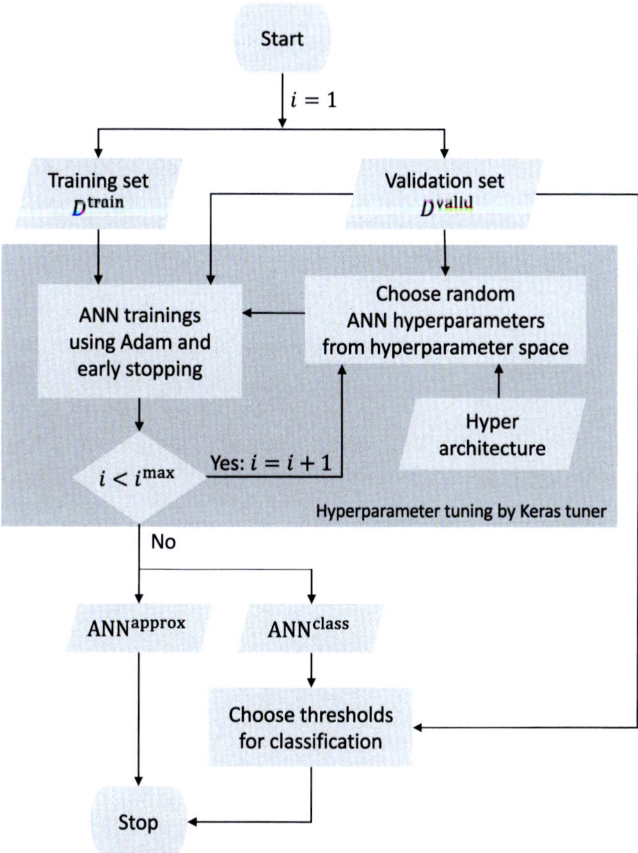

Figure 6.3: Flowchart of the hyperparameter tuning with embedded ANN training. ANN^{approx} and ANN^{class} are based on the same hyper architecture. However, the hyperparameter tuning is done separately such that the best-performing architectures of ANN^{approx} and ANN^{class} can differ.

For both ANNs, we tune the following hyperparameters: the learning rate, the number of layers, the number of neurons per layer, the number of LSTM blocks per LSTM layer, and dropout rates during ANN training (Table 6.1). Dropout is a regularization method we use during ANN training to improve the generalization capability of the ANNs (Srivastava et al., 2014). Here, the dropout rate is the probability of a neuron dropping out before training.

The Keras tuner randomly chooses values for the hyperparameters within the specified hyperparameter space. For new applications, the specified hyperparameter space should offer a wide range of values such that suitable ANN architectures can be found. For orientation, we state the initial hyperparameter space we used in the case study in Table 6.1. A good indicator for a well-chosen hyperparameter space is when the hyperparameters of the best-performing ANNs do not have values close to the minimum or the maximum of the hyperparameter search space. This information also allows further narrowing the parameter limits of the hyperparameter space by performing the hyperparameter tuning repeatedly.

Table 6.1: Hyperparameter search space for the initial hyperparameter tuning in the case study.

Hyperparameter	Minimum	Maximum	Increment
Learning rate α	0.001	0.05	0.002
Number of dense layers	1	5	1
Number of neurons per dense layer	5	200	5
Number of LSTM layers	1	5	1
Number of LSTM blocks per LSTM layer	5	75	5
Number of final dense layers	1	5	1
Number of neurons per final dense layer	5	200	5
Dropout rate in dense layers	0	0.5	0.01
Dropout rate in LSTM	0	0.5	0.01

We set the number of trials to $i^{\max} = 200$. For each trial i, the Keras tuner chooses a new set of hyperparameters and performs a separate training for $\mathrm{ANN}^{\mathrm{approx}}$ and $\mathrm{ANN}^{\mathrm{class}}$. We use the optimizer Adam (Kingma and Ba, 2015) for the ANN training with default settings except for the learning rate α as it is part of the hyperparameter tuning. Adam is an optimization algorithm that can be used instead of classical stochastic gradient descent to update weights of the nets iteratively using training data. We set the maximum number of epochs $e^{\text{max}} = 300$, and an early-stopping patience of 10, meaning that the training ends if the prediction performance on the validation set D^{valid} does not improve for 10 epochs in a row (Graves, 2012). The ANNs are saved at the epoch of their best prediction performance on the validation set D^{valid}. The metrics to measure the prediction performance on the validation set D^{valid} are the sum of squared errors for $\text{ANN}^{\text{approx}}$ and the binary cross-entropy for $\text{ANN}^{\text{class}}$ (Bishop, 2016). After finishing training for all trials i, we select the $\text{ANN}^{\text{approx}}$ with the smallest sum of squared errors and the $\text{ANN}^{\text{class}}$ with the smallest binary cross-entropy on the validation set D^{valid} out of all trials for the application in multi-energy system optimization.

However, before applying $\text{ANN}^{\text{class}}$, we round its output values to binary values. $\text{ANN}^{\text{class}}$ uses the logistic sigmoid activation function in its output layer as recommended for multi-label classification problems (Bishop, 2016). The logistic sigmoid activation function leads to continuous output values between 0 and 1. A threshold value determines if the continuous output values are rounded to 0 or 1. We choose 0.5 as the threshold value such that continuous output values greater or equal 0.5 are rounded to 1, and those below 0.5 are rounded to 0. This value works well for our investigated case studies. However, if such a trivial threshold choice does not work, sophisticated methods to optimize thresholds for multi-label classification can be used (Pillai et al., 2013; Fernández et al., 2018). After the choice of thresholds for $\text{ANN}^{\text{class}}$, we can employ both $\text{ANN}^{\text{approx}}$ and $\text{ANN}^{\text{class}}$ in the decomposition method to boost the operational optimization of multi-energy systems. We implemented the decomposition method in Python 3.8. For the neural nets, we used TensorFlow. For the operational optimization problems, we used the open source toolbox AutoMoG (Kämper et al., 2021d).

6.3 Case Studies

The proposed decomposition method is applied to two case studies. All optimization problems were solved by Gurobi 9.0.0 (Gurobi Optimization, 2020) in standard configuration and a time limit of 10 h using six Intel Skylake 8160 CPUs with a clock speed of 2.1 GHz and 32 GB RAM.

The first case study is a real-world multi-energy system (Section 6.3.1). Here, the operational expenditures ($OPEX$) are normalized due to confidentiality reasons. The second case study is based on a multi-energy system from the literature (Voll et al., 2013) and, in contrast to the first case study, contains a cooling grid and storage units (Section 6.3.2). Finally, an extension of the decomposition method is presented in Section 6.3.3.

6.3.1 Real-world Case Study

The proposed decomposition method is applied to a real-world multi-energy system. The multi-energy system can use the gas grid to purchase gas, a coal supply to purchase coal, and the power grid to purchase and sell electricity. The multi-energy system has to fulfill demands for steam on two pressure levels, power, and compressed-air. The components of the multi-energy system are multiple coal-driven and gas-driven boilers, a gas turbine combined with a heat recovery boiler, and multiple steam turbines. Starting up any component causes start costs. The coal-driven boilers and some steam turbines are restricted by minimum up- and downtimes. Additionally, the operation of the coal-driven boilers is restricted by ramping constraints.

We apply the decomposition method to solve the operational optimization of the multi-energy system for one week with hourly resolution. To apply the proposed decomposition method, we perform the data acquisition for the ANN training as described in Section 6.2.2. The historical time series of the demands and the prices for power, gas, and coal are available for a full year (8760 time steps). With these time series, we generate 30 synthetic years of all time series following Klöckl and Papaefthymiou (2010), leading to 806 samples of two weeks each (including the historical time series). We solve the 806 operational optimizations to create the input - target pairs. We split the samples into a training set D^{train} (70 % of the samples), a validation set D^{valid} (15 % of the samples), and a test set D^{test} (15 % of the samples). With the training set D^{train} and the validation set D^{valid}, we perform the hyperparameter tuning and training of the ANNs as described in Section 6.2.3. The results of the hyperparameter tuning and training are shown in Appendix A. For hyperparameter tuning and training of the ANNs, we have to choose the specific variables that the ANNs predict. Tests on the real-world case study indicated that predicting the loads and on/off states of all components leads to higher increases in $OPEX$ than predicting only the time-coupled variables. Thus, for the real-world case study, $\text{ANN}^{\text{class}}$ predicts the on/off states of the steam turbines and the coal-driven boilers since these components are restricted by minimum up- and downtimes. $\text{ANN}^{\text{approx}}$ predicts the load of the coal-driven boilers since these components are additionally restricted by ramping constraints. Finally, we solve the operational optimization problems for the 121 samples of the test set D^{test}.

We apply two benchmarks to solve the operational optimization problems of the given multi-energy system: First, we solve the operational optimization problems with Perfect Foresight, meaning that we solve all 168 time steps in one optimization problem. Second, we solve the operational optimization problems with single-time-step optimizations to capture the effect of the ANN predictions.

In the following, we present the results of the operational optimization for the decomposition method and the benchmarks. In Section A, we state the results of the hyperparameter tuning and training of the ANNs.

First, we analyze the computation times of the operational optimization. Figure 6.4 shows a histogram of the computation times for the 121 operational optimization problems of one week with hourly resolution from the test set D^{test}.

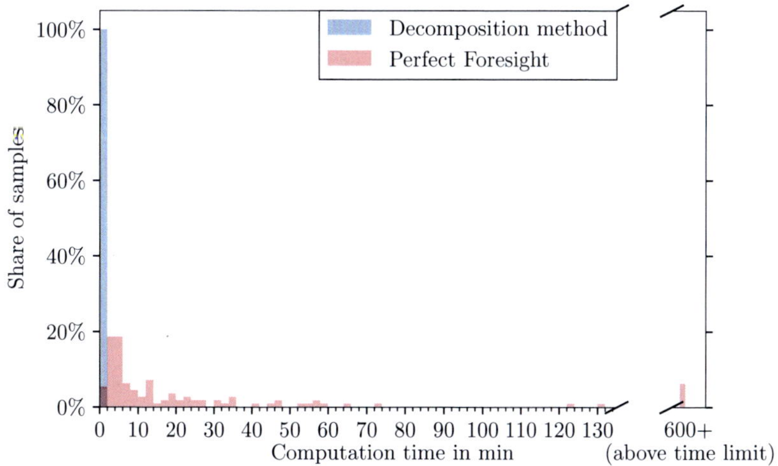

Figure 6.4: Histogram of the computation times for the decomposition method and Perfect Foresight in the real-world case study. The decomposition method (blue) solves all operational optimization problems in less than 2 min such that all bars collapse in one point.

The proposed decomposition method solves 100 % of the operational optimization problems in less than 2 min. The average computation time for the operational optimization of one week is 81 s with a standard deviation of $\sigma = 3$ s. Even more, the average cumulative solver time for one week is only 8 s ($\sigma = 1$ s). The remaining computation time is mainly needed to transfer the data between single-time-step optimizations and ANN predictions. The results show that the decomposition method solves the operational optimization in a reliably short computation time.

In contrast, the computation times of Perfect Foresight are unpredictable and range from less than 2 min up to the time limit of 10 h (Figure 6.4). The average computation time is 3204 s with a standard deviation of $\sigma = 8600$ s. Thus, the proposed decomposition method reduces the average computation time by a factor of 40 and the worst computation time even by a factor 440. For Perfect Foresight, the total computation time and the solver time hardly differ. As expected, the benchmark using single-time-step optimizations without ANN predictions is faster than the decomposition method, solving an operational optimization problem of one week in 34 s on average with a standard deviation of $\sigma = 3$ s. The average cumulative solver time is 8 s, the same as for the decomposition method. Thus, the decomposition method is almost as fast as a single-time-step computation.

Concerning the solution quality, the benchmark using single-time-step optimizations without ANN predictions increases the $OPEX$ by 4.6 % on average (Figure 6.5). Single-time-step optimization shows distinct diurnal periodicities due to daily fluctuations of energy prices and demands. At certain time steps, the $OPEX_t$ is even smaller than the $OPEX_t$ of Perfect Foresight. However, the single-time-step optimization always pays for those time steps with a high increase of the $OPEX_t$ a few time steps later such that the total $OPEX$ for the entire week can never be lower than the $OPEX$ of Perfect Foresight.

For the decomposition method, the $OPEX$ increases only 1.4 % on average compared to Perfect Foresight. The increase in $OPEX_t$ for the decomposition method is systematically lower than for single-time-step optimization. The diurnal periodicities are significantly reduced for the decomposition method because the ANN predictions incorporate information about the price and demand patterns. The upper limit of the 90 % confidence intervals of the decomposition method is on the same level as the average $OPEX_t$ increase of the single-time-step optimization. The confidence interval shows that 90 % of the solutions from the decomposition method are at least as good as the average solution of the single-time-step optimization. These results demonstrate that the ANN predictions significantly improve the solution quality of single-time-step optimizations.

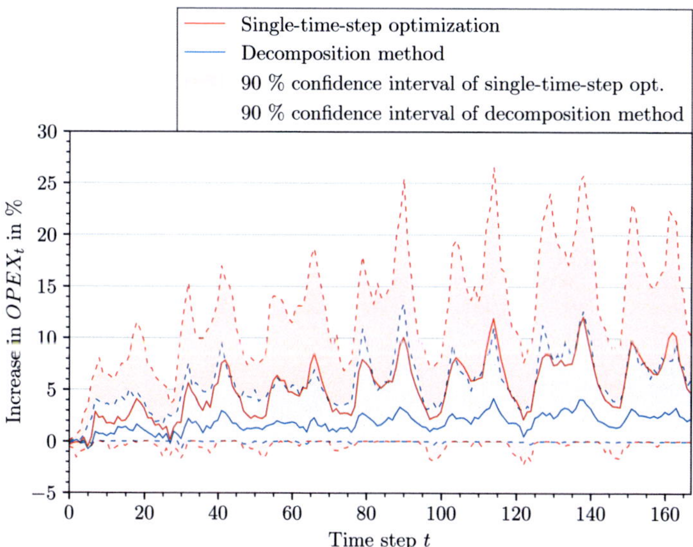

Figure 6.5: Increase in $OPEX_t$ per time step t in % compared to Perfect Foresight for the decomposition method (blue) and single-time-step optimization (red).

6.3.2 Literature Case Study

We apply the proposed decomposition method to another multi-energy system based on the literature example from Voll et al. (2013) (Figure 6.6). For this literature case study, we use the part-load curves of the gas boilers, absorption chiller, compression chillers, and combined-heat-and-power (CHP) units from Voll et al. (2013), and add a coal boiler, an electrode boiler, turbines, valves and storage units. Furthermore, we establish minimum loads, minimum up- and downtimes, ramping constraints, and start costs for the components. The literature case study is more complex compared to the real-world case study and shows the transferability of the decomposition method to operational optimization problems of other multi-energy systems.

As in the real-world case study, the multi-energy system in the literature case study can use the gas grid to purchase gas, a coal supply to purchase coal, and the power grid to purchase and sell electricity. However, the multi-energy system has to fulfill two heat demands on different temperature levels, a cooling demand, and a power demand. Furthermore, the multi-energy system contains additional components compared to

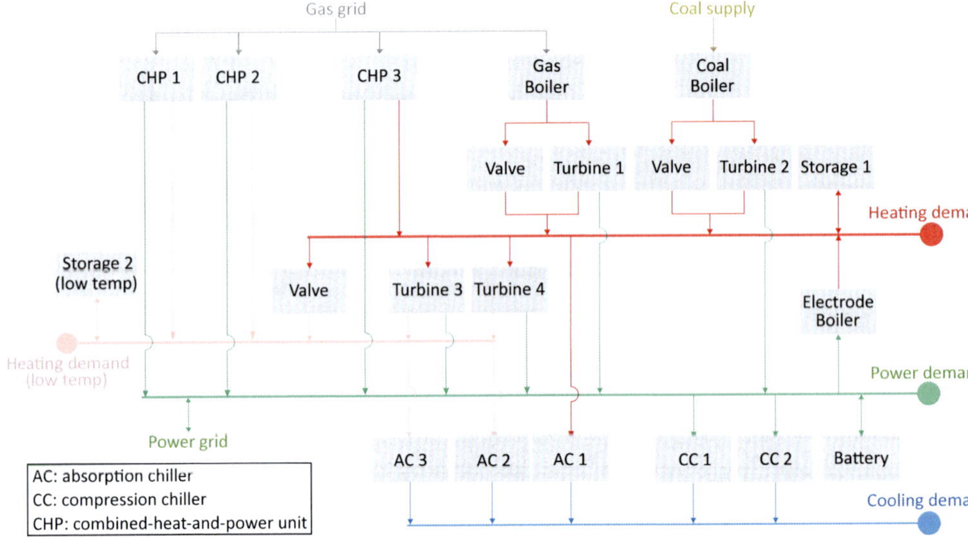

Figure 6.6: Multi-energy system in the literature case study.

the real-world case study, such as compression chillers, absorption chillers, an electrode boiler, and energy storage units. The operation of nearly all units is restricted by minimum loads, minimum up- and downtimes, and ramping constraints. The details of the components in the multi-energy system are given in Table 6.2.

As in the previous case study, we apply the decomposition method to solve the operational optimization of the multi-energy system for one week with hourly resolution. In this case study, we generate 25 synthetic years of all time series, leading to 650 samples of two weeks each and 98 operational optimization problems in the test set D^{test}. For the literature case study, the number of time-coupled variables is higher since all components are restricted by minimum up- and downtimes, ramping constraints, and start costs. The ANNs predict the on/off states of all components listed in Table 6.2, and the loads of the three CHP units, the three absorption chillers, the level of the three storage units, and the exchange of electricity with the grid. We apply the same two benchmark approaches to solve the operational optimization problems in the test set D^{test}.

The strong operational restrictions in the multi-energy system of the literature case study raise a problem that did not appear in the real-world case study: infeasible

Table 6.2: Details of the components in the multi-energy system.

Component	Nominal output (Capacity)	Maximum ramp	Minimum part load	Start cost	Minimum up- and downtime
CHP 1	2.5 MW^{th}	1.25 MW/h	1.25 MW^{th}	1000 €	12 h
CHP 2	2.3 MW^{th}	1.15 MW/h	1.15 MW^{th}	1000 €	12 h
CHP 3	4.6 MW^{th}	2.4 MW/h	2.3 MW^{th}	1250 €	12 h
Gas Boiler	11.3 MW	3.7 MW/h	2.26 MW	2000 €	12 h
Coal Boiler	14.7 MW	2.2 MW/h	2.94 MW	3000 €	12 h
Electrode Boiler	8 MW	8 MW/h	2 MW	0 €	0 h
Turbine 1,2,3,4	1.45 MW^{el}	1.45 MW/h	0.6 MW^{el}	500 €	3 h
AC 1	6.5 MW	3.3 MW/h	1.3 MW	650 €	3 h
AC 2	3.5 MW	1.9 MW/h	0.7 MW	350 €	3 h
AC 3	2 MW	1.05 MW/h	0.4 MW	200 €	3 h
CC 1	2.5 MW	2.5 MW/h	0.5 MW	250 €	3 h
CC 2	1.6 MW	1.6 MW/h	0.32 MW	160 €	3 h
Storage 1	10 MW (50 MW h)	10 MW/h	0 MW	0 €	0 h
Storage 2	6 MW (25 MW h)	6 MW/h	0 MW	0 €	0 h
Battery	0.75 MW (5 MW h)	0.75 MW/h	0 MW	0 €	0 h

optimization problems due to the inertia of the components. Thus, we discuss the infeasibilities before we analyze computation times and solution quality.

The original problems are feasible, demonstrated by the fact that Perfect Foresight provides a solution for all problems in the test set. In contrast, single-time-step optimization only provides a feasible solution for 30 of the 98 problems in the test set. In the 68 remaining problems, infeasible subproblems arise because the multi-energy system cannot change its system operation from the last and now fixed time step to the current time step. All observed infeasibilities are due to the highly restricted components. The components cannot react to high fluctuations in energy demands such that energy balances become infeasible. Mainly, the infeasibilities arise in the energy balance of the cooling grid. The lack of cooling storage, the minimum up- and downtimes, and the ramping constraints of the components that supply cooling lead to an overall inflexible cooling supply. This inflexible cooling supply and the lack of any future information quickly lead to infeasibilities in single-time-step optimization. In contrast, the decomposition method solves 86 of the 98 problems in the test set, showing that the ANN predictions strongly help to avoid infeasibilities. Even in the 12 infeasible problems, the ANN predictions for the on/off states of the components are exactly the same as the decisions in the perfect-foresight solutions. The ANN predictions for the component loads are very similar to the perfect-foresight solutions but not exactly the same. The deviations are between 0.001 and 0.03 MW. These small deviations are still sufficient such that no other component can compensate them in these 12 problems due to the highly restrictive combination of maximum ramps, minimum up- and downtimes, and minimum part load. In practice, these minor deviations are most likely no problem, but the mathematical model has no margin on the hard constraints. However, for other case studies, the share of infeasible optimization subproblems may increase with more fluctuating energy demands or if renewable energy resources are paired with energy system components that are even less flexible than those in the presented literature case study. The decomposition method may then be extended to overcome the infeasibility problems if the ANNs provide the optimal values for the binary variables (see Section 6.3.3). The following analysis of computation times and solution quality refers only to the feasible problems of the respective solution approaches.

Figure 6.7 shows a histogram of the computation times for the 98 operational optimization problems of one week with hourly resolution from the test set.

As in the real-world case study, the decomposition method solves 100 % of the operational optimization problems in less than 2 min. Here, the average computation time for the operational optimization of one week is 103 s with a standard deviation of $\sigma = 4$ s. The computation time is slightly higher compared to the real-world example since the amount of data transferred between single-time-step optimizations and ANN predictions is larger. Still, the decomposition method solves the operational optimization in a reliably short computation time despite the significantly more complex multi-energy system. In contrast, Perfect Foresight reaches the time limit of 10 h for all samples before reaching the optimality gap. The remaining optimality gap after reaching the time limit is 1.3 % on average.

Figure 6.8 shows the speed-up factor realized by the decomposition method compared to Perfect Foresight for the 98 samples from the test set. The average speed-up factor is 375, the minimum is 317, and the maximum is 821.

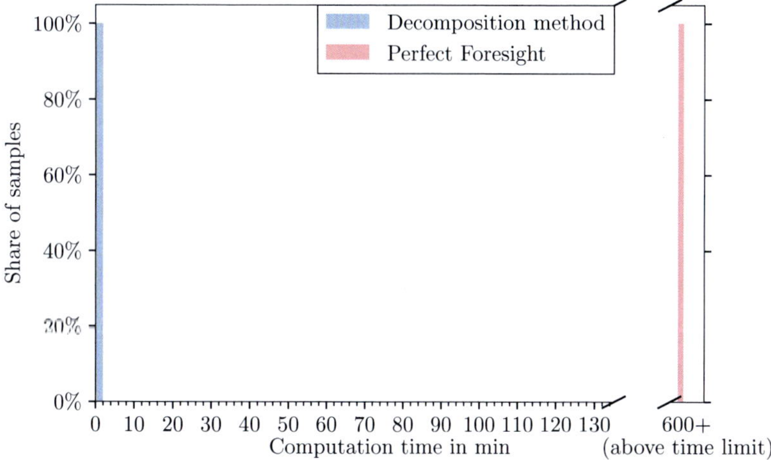

Figure 6.7: Histogram of the computation times for the decomposition method and Perfect Foresight in the literature case study. The decomposition method (blue) solves all operational optimization problems in less than 2 min. Perfect foresight reaches the time limit in all operational optimization problems.

The benchmark using single-time-step optimizations without ANN predictions solves one operational optimization problem in 43 s on average with a standard deviation of $\sigma = 3\,\text{s}$. Note that this computation time is the average of only the 30 problems in the test set for which the benchmark provided a solution. In summary, the literature case study shows that the computation time of the decomposition method hardly increases with the complexity of the multi-energy system although the computation time of Perfect Foresight drastically increases. As a result, the decomposition method realizes high speed-up factors for all samples in the literature case study.

In the literature case study, the average increase in $OPEX$ is about 4.9 % for the decomposition method compared to the best solutions found by Perfect Foresight within the time limit. The solutions of the benchmark using single-time-step optimizations without ANN predictions show a significantly higher increase in $OPEX$ of about 65 % on average (Figure 6.9). Similar to the real-world case study, the increase in $OPEX_t$ for the decomposition method is systematically lower than for single-time-step optimization. The differences in solution quality between the decomposition method

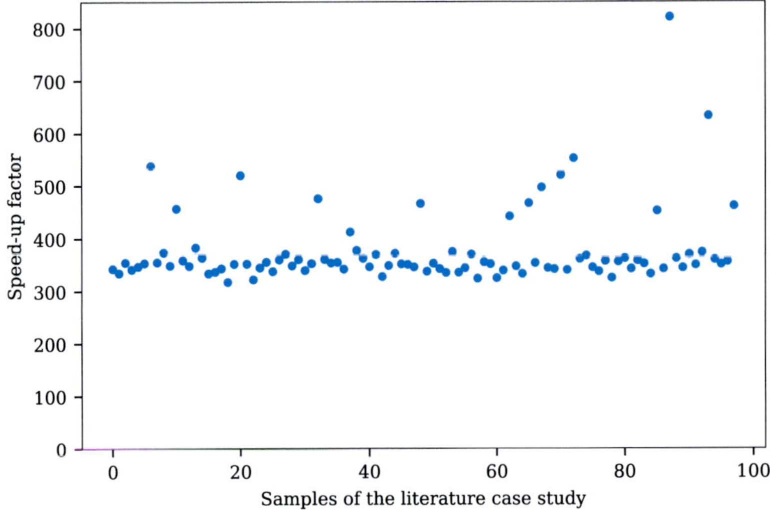

Figure 6.8: Speed-up factor realized by the decomposition method compared to Perfect Foresight for the 98 samples from the test set of the literature case study.

and single-time-step optimization are even larger in the literature case study. In Figure 6.9, single-time-step optimization shows high peaks and fluctuations, whereas the decomposition method shows a slight and smooth increase in $OPEX$ over the whole week.

In the following, we compare the use of the storage units for all approaches. Figure 6.10 shows the storage levels of the three storage units over time for one week from the test set. For all other problems in the test set, the results are qualitatively similar for all storage units and solution approaches. The levels of the storage units are nearly the same for the decomposition method and Perfect Foresight. In contrast, the benchmark using single-time-step optimizations without ANN predictions has no information about the future. Therefore, the use of all storage units strongly differs for single-time-step optimization compared to the decomposition method and Perfect Foresight (Figure 6.10). Single-time-step optimization does not use the battery because it is always more beneficial to use the electricity for the demand or sell it instead of loading the battery, when only one time step is considered.

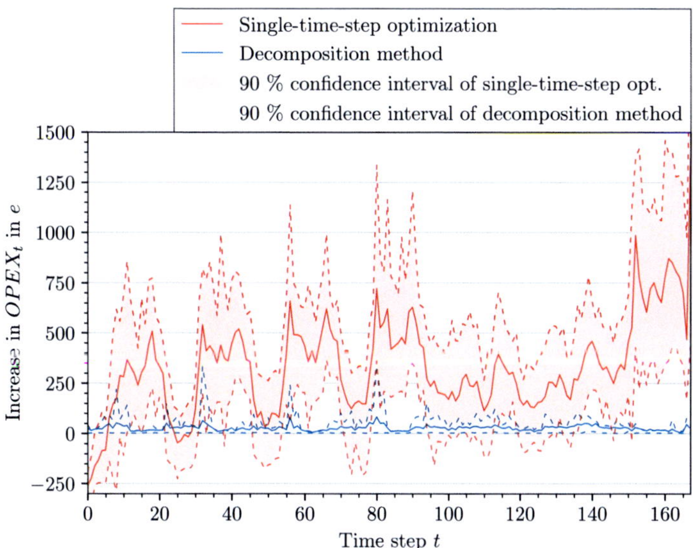

Figure 6.9: Increase in $OPEX_t$ per time step t in e compared to Perfect Foresight for the decomposition method (blue) and single-time-step optimization (red).

Over all time steps of 86 feasible problems in the test set, the storage levels in the decomposition method show an average deviation of 0.39 MW h for Storage 1, 0.05 MW h for Storage 2, and 0.01 MW h for the battery compared to the storage levels in Perfect Foresight. These small deviations show that the decomposition method can use the storage units as weekly storage although performing single-time-step optimizations. In contrast, the benchmark using single-time-step optimizations without ANN predictions shows significantly higher average deviations in the storage levels of 19.7 MW h for Storage 1, 9.4 MW h for Storage 2, and 2.4 MW h for the battery compared to Perfect Foresight. These results of the benchmark using single-time-step optimizations confirm that the use of the storage units over one whole week is not straightforward.

Altogether, the literature case study reveals that the decomposition method can be transferred to multi-energy systems incorporating storage units and many strongly restricted components. The decomposition method significantly reduces the computation time in both case studies to a reliably short time below 2 min. Still, the

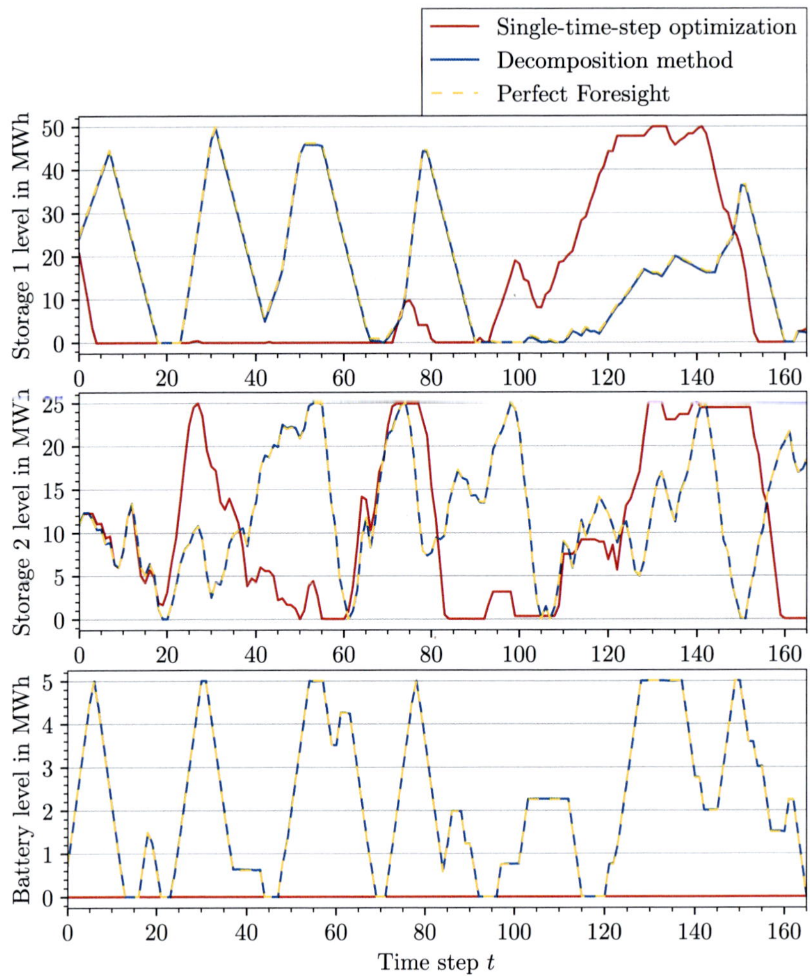

Figure 6.10: Levels of storage 1 (top), storage 2 (middle), and the battery (bottom) over all time steps of one exemplary week from the test set.

decomposition method shows a good solution quality in both case studies. The use of the storage units proves that the decomposition method makes long-term decisions in single-time-step optimizations.

6.3.3 Extension of the decomposition method

The reliably short computation time allows to extend the decomposition method by a Perfect-Foresight optimization with fixed binary variables. We demonstrate the optional extension of the decomposition method as indicated in Section 6.2.1 using the literature case study. In this extension, we extract the values of the predicted binary variables (on/off states of all components in Table 6.2) from the solution provided by the decomposition method. We transfer these values of the binary variables into the original optimization problem and fix them. Subsequently, we solve the original problem with the fixed binary variables using Perfect Foresight. In the following, we refer to this approach as the extended decomposition method.

We apply the extended decomposition method to the literature case study. Figure 6.11 shows the remaining optimality gap over time for Perfect Foresight without fixing any variables and the extended decomposition method. For Perfect Foresight without fixing any variables, the time corresponds to the pure solver time up to the time limit of 36.000 s (10 h). As reported in Chapter 6.3.2, Perfect Foresight does not reach the standard optimality gap of 10^{-4} within the time limit in any of the 98 problems from the test set. The average optimality gap after reaching the time limit is 1.3 %. For the extended decomposition method, two times are shown (Figure 6.11): The blue lines correspond to the total computation time of the extended decomposition method. The black lines correspond to the cumulated solver time of the extended decomposition method needed to reach the standard optimality gap of 10^{-4}.

Even when considering the total time of the extended decomposition method, the arithmetic average time to reach the optimality gap is 170 s with a standard deviation of 233 s. The longest observed computation time to reach the optimality gap with the extended decomposition method is 2176 s (roughly 36 min). This sample is the only outlier as the second-longest time observed is 158 s, and thus, already below the average computation time.

The results show that the extended decomposition method drastically reduces the computation time for a Perfect-Foresight optimization. However, the optimization problems solved by the extended decomposition method and Perfect Foresight are not the same as we fix binary variables in the extended decomposition method. Therefore, the extended decomposition method is still a primal heuristic for the original opti-

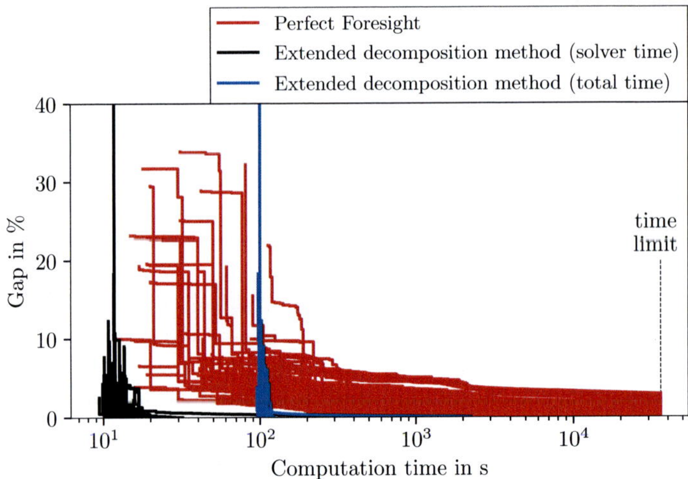

Figure 6.11: Optimality gap over time for Perfect Foresight without fixing any variables (red) and the extended decomposition method (blue for total computation time, black for cumulative solver time). The blue lines correspond to the maximum time needed to apply the extended decomposition method. The black lines correspond to the minimum time to be approached by a more efficient data handling between the time steps.

mization problem. Nonetheless, the increase in $OPEX$ compared to Perfect Foresight is $0.1\,\%$ on average. Thus, the extension enables to improve solution quality of the decomposition method and still provides a solution in a short time. However, the reliability of the computation time decreases.

The extended decomposition method is a promising approach when many binary variables can be taken from the solutions of the decomposition method and fixed to reduce the original operational optimization problems. Furthermore, the extended decomposition method may help in case studies with notable infeasibility problems, provided that the ANNs still predict the optimal values for the binary variables. In those cases, the strategy could be to relax the energy balances in the single-time-step optimizations to avoid infeasibilities. Thereby, the ANNs predict the binary variables for each time step. Afterward, we fix these binary variables and solve the original problem without any relaxation, as the extended decomposition method suggests.

However, this strategy only works with high accuracy if the ANNs provide optimal values for the binary variables.

6.4 Conclusion

A decomposition method is proposed to boost the operational optimization of multi-energy systems. The method decomposes the operational optimization problem into single-time-step optimizations. The method incorporates ANN predictions into the single-time-step optimizations to improve the solution quality. The ANNs are trained on long-term solutions and predict binary and continuous decision variables. Thereby, long-term effects can be considered in the operational optimization of a single time step.

In the case study of a real-world multi-energy system, the decomposition method solves the operational optimization in less than 2 min for every problem instance. In contrast to Perfect Foresight, the computation time of the decomposition method is independent of the problem instance, and the decomposition method provides the solution about 40 times faster than Perfect Foresight on average, even more than 400 times faster in the eight problem instances that reach the time limit of 10 h. On average, the increase in $OPEX$ for the decomposition method is about 1.4 % compared to 4.6 % for single-time-step optimizations without ANN predictions.

The second case study from literature shows that the decomposition method can be applied to even more complex multi-energy systems that include storage units. Still, the computation time of the decomposition method remains less than 2 min for every problem instance, although the computation time of Perfect Foresight drastically increases. For every problem instance, Perfect Foresight reaches the time limit of 10 h before reaching the optimality gap. The solution quality of the decomposition method is significantly better than single-time-step optimizations without ANN predictions (4.9 % vs. 65 % increase in $OPEX$ compared to Perfect Foresight). The decomposition method uses storage units as weekly storage, although it decomposes the original problem into single-time-step optimizations. Thereby, the decomposition method shows its ability to make long-term decisions in single-time-step optimizations.

In the literature case study, infeasibilities arise due to highly restricted components. The benchmark using single-time-step optimizations without ANN predictions only yields a feasible solution for 30 of 98 problem instances. However, the proposed decomposition method delivers a feasible solution for 86 of 98 problem instances, showing that the ANN predictions help to avoid infeasibilities although using single-time-step optimizations.

When applying the decomposition method to other case studies, the specific variables to be predicted have to be chosen. In the presented case studies, predicting the time-coupled variables of the multi-energy system models leads to lower $OPEX$ than predicting all variables. Thus, we suggest predicting all time-coupled variables first and then trying other sets of variables to be predicted to see which set works best. Special attention should be paid to the most restricted components.

In summary, the presented case studies show that the decomposition method solves operational optimization problems in a reliably short time and that ANN predictions improve the solution quality of single-time-step optimizations. The results motivate deepening the combination of optimization and machine learning.

CHAPTER 7

Summary, Conclusions, and Future Perspectives

The goal of this thesis is to provide methods that enable a broader application of mathematical optimization in real-world problems. In the following, we summarize the thesis and draw the main conclusions in Section 7.1. In Section 7.2, we discuss possible perspectives on future research directions in the field of multi-energy system modeling and optimization.

7.1 Summary and Conclusions

The increasing amount of measured data in multi-energy systems provides possibilities to develop new methods that use the data to meet today's challenges. A major challenge of today is the optimal operation of multi-energy systems while the energy sector transforms towards sustainability to mitigate climate change. Mathematical optimization is well suited for optimally operating multi-energy systems.

Operational optimization requires a mathematical model of the multi-energy systems. However, generating a mathematical model of a multi-energy system is tedious and, thus, time-consuming and expensive. Furthermore, the increasing share of renewable energy resources and the associated volatility forces operational optimization to be solved repeatedly and fast. Solving operational optimization of multi-energy systems fast is challenging due to the inherent complexity of the systems leading to computational complexity. To meet these challenges, we present methods in this thesis that use data of multi-energy systems to automatically generate mathematical models as well as accelerate operational optimization. The presented methods cover the complete process from measured data to optimal operation of a multi-energy system.

The first challenge in this process is the data-driven model generation. Operational optimization desires accurate and computationally efficient models. In general, the data-driven generation of accurate and computationally efficient models yields complex optimization problems itself (Section 3.1). In Chapter 3, we propose the AutoMoG method for automated data-driven model generation of multi-energy systems. AutoMoG decomposes the data-driven model generation problem of a multi-energy system into one model generation problem for each component. Still, the error of model generation is evaluated for the overall multi-energy system. The solution provided by AutoMoG is an MILP model of the multi-energy system. MILP models allow for efficient optimization by applying state-of-the-art solvers. In the presented case study, AutoMoG needs significantly fewer piecewise-affine sections to reach an allowed model error compared to the commonly employed independent modeling of each component. As a result, the less complex model of AutoMoG solves the operational optimization more than 50 times faster on average without a loss in solution quality. The results of the case study show that each component's modeling error should be seen in the context of the multi-energy system model.

To further push the applicability of mathematical optimization, the AutoMoG method is extended to AutoMoG 3D in Chapter 4. In contrast to the original AutoMoG method, AutoMoG 3D is able to generate models of systems that contain components with multiple independent variables by applying hinging-hyperplane trees. The case study of an industrial real-world pump system shows that AutoMoG 3D is even applicable to synthesis problems. The model provided by AutoMoG 3D solves the synthesis problem ten times faster than the generalized-convex-combination model without any loss in the solution quality. AutoMoG and AutoMoG 3D are not limited to generating models of multi-energy systems but can generate models of any engineering system. AutoMoG and AutoMoG 3D drastically decrease the effort for data-driven model generation and optimization. Thus, the methods contribute to the goal of this thesis to enable a broader application of mathematical optimization in real-world problems. We released a Python version of the code open-source for both methods to enable a straightforward application (Kämper et al., 2021d).

Once the model of the multi-energy system is generated, the following challenge is the fast solution of the operational optimization with this model. A common and easily applicable approach to solving operational optimization fast is the Rolling-Horizon method. However, in general, the quality of Rolling-Horizon solutions differs between problem instances and strongly depends on the chosen values of the step size and foresight. For this reason, we propose the Adaptive-Rolling-Horizon approach in Chapter 5. The approach adaptively determines an individual foresight for each subproblem based on knowledge about the multi-energy system and its time-dependent parameters (e.g., time series of energy prices). Thereby, the Adaptive Rolling Horizon reduces the computation time of operational optimization while ensuring correct start-ups and shutdowns of the components. As a result, the Adaptive Rolling Horizon reaches excellent solution quality. In a real-world case study, the Adaptive Rolling Horizon reduces the computation time by a factor of 150 on average compared to Perfect Foresight. Still, the approach provides solutions near to the globally optimal solution, increasing $OPEX$ by less than 0.1 % on average.

While the Adaptive Rolling Horizon significantly reduces the computation time, the method still cannot guarantee a reliably short computation time independently of the problem instance. For this reason, we present a decomposition method that applies artificial neural nets (ANNs) for the operational optimization of multi-energy systems in Chapter 6. The method decomposes the operational optimization problem into single-time-step optimizations. To improve the solution quality, the method incorporates ANN predictions in the single-time-step optimizations. The ANNs are trained on long-term solutions and predict decision variables of the operational optimization to consider long-term effects in a single time step. The decomposition method guarantees a reliably short computation time and solves all operational optimization problems of one week in less than 2 min for a real-world case study and more complex literature case study. In contrast, the computation times of Perfect Foresight range from less than 2 min up to the time limit of 10 h in the real-world case study. In the literature case study, Perfect foresight reaches the time limit of 10 h for every problem instance. In the real-world case study, the solution quality of the decomposition method in terms of $OPEX$ is about 1.4 % worse than Perfect Foresight. Still, comparing the solution quality to single-time-step optimizations without ANN predictions (about 4.6 % worse than Perfect Foresight) shows that the ANNs systematically improve the solution quality. In the literature case study, the increase in $OPEX$ compared to Perfect Foresight is 4.9 % for the decomposition method. However, for single-time-step optimizations without ANN predictions, the increase in $OPEX$ is significantly higher at 65 %. The decomposition method uses storage units as weekly storage, although it decomposes the original problem into single-time-step optimizations. In summary, the results show that the decomposition method provides high-quality solutions for the operational optimization of multi-energy systems in reliably short times.

The presented methods facilitate and accelerate the data-driven model generation and operational optimization of multi-energy systems. Thereby, the methods enable a broader application of mathematical optimization in real-world problems. Nonetheless, there are still challenges to overcome in the field of data-driven modeling and optimization of multi-energy systems. We sketch possible future perspectives in the next section.

7.2 Future Perspectives

In this section, we discuss possible future perspectives in data-driven multi-energy system modeling and optimization. The topics are ordered chronologically along the process from measured data to optimal operation of a multi-energy system and provide future perspectives of the steps involved in the process.

Data preparation. In this thesis, we start with prepared data and then apply automated modeling methods. However, in reality, the available raw data often lacks sufficient quality and consistency due to, e.g., missing or defective measuring devices, or inconsistent measuring methods. The effort is immense to prepare the data such that it can be used in subsequent modeling methods. For this reason, methods are highly desired that enable fast or even automated data preparation. Such methods could be based on approaches like, e.g., data imputation (Lin and Tsai, 2020) to close gaps in incomplete data sets or data reconciliation (Wahab et al., 2021) to correct or reduce measurement errors. Methods that automate data preparation can be combined with data-driven methods that use the prepared data, such as the methods presented in this thesis.

Determining a sufficient modeling detail. Which model is the better model? When using models to optimize real systems, we should be aware that there is always a mismatch between model and real system. Intuitively, a model with a smaller mismatch between the model and the real system would be seen as *better* since it represents the real system more accurately. For this reason, a modeler typically selects a model based on accuracy measures. In practice, a higher accuracy only has value if it leads to better decisions. Thus, there is no reason to use a detailed and probably slow model when a simple and fast model leads to the same decisions for the real system. Thus, new metrics are needed to reach a sufficient modeling detail by raising the question: does a more detailed model lead to better decisions for the real system, or does it just reduce the mismatch between the model and the real system?

Refinement criterion in automated model generation. AutoMoG and AutoMoG 3D use cost-based weighting factors to decide which component to refine and when to stop model refinement. The cost-based weighting factors show good performance and low sensitivity in the presented case study. However, choosing cost-based weighting factors is not straightforward when considering other piecewise-affine approximations, e.g., for nonlinear investment-cost curves of components in synthesis problems. Therefore, it might be beneficial to consider other options than cost-based weighting factors. The iterative process of model refinement in the AutoMoG methods allows considering the actual optimization during model generation. Thus, the objective function could be evaluated for each possible model refinement to decide which component to refine and when to stop model refinement. A representative optimization problem could then be solved for each possible model refinement. In that case, the refinement with the highest impact on the objective function would be chosen. This approach allows to include other piecewise-affine approximations than part load easily.

Problem-specific predictions of artificial neural nets in optimization. The decomposition method from Chapter 6 uses the same ANNs for solving the operational optimization of every problem instance. A common challenge for ANN training is that the training samples have to be sufficiently similar to expected future problem instances for satisfying results. To get better predictions for the decision variables in an optimization problem, the ANNs could be tailored to the specific problem instance. Tailoring the ANNs could be reached by first clustering the training samples and then training ANNs for each cluster. Before solving an optimization problem, the similarity of the problem instance with the clusters would be checked, and the corresponding ANNs would be used for the predictions.

Considering seasonal effects in daily optimizations. Multi-energy systems are often subject to seasonal constraints such as emission limits, seasonal storage, network-connection fees, or peak-power prices. Methods are available to solve seasonal operational optimization. However, the computation times of these methods prevent an application in daily optimizations. Thus, it is important to consider seasonal effects in daily optimizations to track the long-term optimum. One possible approach is, e.g., to solve multiple one-year optimizations to develop operational strategies for seasonal storage. The one-year optimizations could be performed, e.g., weekly, to update the operational strategies for seasonal storage. The daily minima and maxima of these strategies could be employed as lower and upper bounds of the storage level for the daily multi-energy system optimization.

The discussed future perspectives support the applicability of mathematical optimization by seizing opportunities provided by the era of big data. In particular, combining methods that automatically prepare data with data-driven methods is the next step towards a holistic big-data concept in many research fields. Pursuing a holistic big-data concept should be part of almost every research field since:

'The world is one big data problem.'
Andrew McAfee

Appendices

Appendix A

Results of hyperparameter tuning and training

The hyperparameters of the ANNs used for the real-world case study are shown in Table A.1. We performed three hyperparameter tunings with 200 trials ($i^{\max} = 200$) each, mainly because we initially overestimated the learning rate α. After the third tuning, the hyperparameters of the best-performing ANNs do not lie on the edges of their allowed range.

Table A.1: Results of the hyperparameter tuning for $\mathrm{ANN}^{\mathrm{approx}}$ and $\mathrm{ANN}^{\mathrm{class}}$. The given hyperparameters correspond to the ANN architectures that performed best on the validation set D^{valid}.

Hyperparameter	$\mathrm{ANN}^{\mathrm{approx}}$	$\mathrm{ANN}^{\mathrm{class}}$
Learning rate α	0.0016	0.001
Number of dense layers	4	1
Number of neurons per dense layer	110	120
Number of LSTM layers	3	2
Number of LSTM blocks per LSTM layer	40	30
Number of final dense layers	1	4
Number of neurons per final dense layer	130	180
Dropout rate in dense layers	0.01	0.065
Dropout rate in LSTM	0.01	0.01

APPENDIX B

Publications and Student Theses

List of publications

Journal publications

Kämper, A., Holtwerth, A., Leenders, L., and Bardow, A. (2021b). AutoMoG 3D: Automated Data-Driven Model Generation of Multi-Energy Systems Using Hinging Hyperplanes. *Frontiers in Energy Research*, 9, 719658.

Kämper, A., Leenders, L., Bahl, B., and Bardow, A. (2021c). AutoMoG: Automated data-driven Model Generation of multi-energy systems using piecewise-linear regression. *Computers & Chemical Engineering*, 145, 107162.

Reinert, C., Schellhas, L., Mannhardt, J., Shu, D., Kämper, A., Baumgärtner, N., Deutz, S., and Bardow, A. (2022). SecMOD: An Open-Source Modular Framework Combining Multi-Sector System Optimization and Life-Cycle Assessment. *Frontiers in Energy Research*, 10, 884525.

Kämper, A., Delorme, R., Leenders, L., and Bardow, A. (2023). Boosting Operational Optimization of Multi-Energy Systems by Artificial Neural Nets. *Computers & Chemical Engineering*, 173, 108208.

Peer-reviewed conference publications

Seele, H., Steufmehl, F., Hagedorn, D., Kämper, A., Reinert, C., and von der Aßen, N. (2022). Optimization-based scheduling of heat-integrated multipurpose batch plants using a discrete temperature grid. *Submitted to ECOS 2022 - 35th International Conference on Efficiency, Cost, Optimization, Simulation and Environmental Impact of Energy Systems*, Copenhagen, Denmark.

Kämper, A., Geers, P., Leenders, L., and Bardow, A. (2021a). Adaptive Rolling Horizon for operational optimization of multi-energy systems. In *Proceedings of the ECOS 2021 - 34th International Conference on Efficiency, Cost, Optimization, Simulation and Environmental Impact of Energy Systems (ECOS 2021)*, Taormina, Italy.

Hoffrogge, D., Schulze Balhorn, L., Seele, H., Kämper, A., and von der Aßen, N. (2021). HENDling: Simultaneous Heat Exchanger Network Design and Scheduling for Batch Processes. In *Proceedings of the ECOS 2021 - 34th International Conference on Efficiency, Cost, Optimization, Simulation and Environmental Impact of Energy Systems (ECOS 2021)*, Taormina, Italy.

Hüttermann, A., Leenders, L., Bahl, B., and Bardow, A. (2019). Automated data-driven model generation of energy systems using piecewise linear regression. In Stanek, W., Gładysz, P., Werle, S., Adamczyk, W., editors, *ECOS 2019 - Proceedings of the 32nd International Conference on Efficiency, Cost, Optimization, Simulation and Environmental Impact of Energy Systems*, pages 327-338.

Further conference contributions

Hoffrogge, D., Schulze Balhorn, L., Kämper, A., Baumgärtner, N., Mayer, F., and von der Aßen, N. (2021). HENDling: Simultaneous Heat Exchanger Network Design and Scheduling for Batch Processes. In *Computer Aided Process Engineering FORUM 2020*, online.

Hüttermann, A., Leenders, L., Bahl, B., and Bardow, A. (2019). Automated data-driven model generation of energy systems using piecewise linear regression. *Jahrestreffen Forschungsnetzwerk Energiesystemanalyse*, Aachen, Germany.

Student theses supervised during this work

The following student theses are described partially in this work:

Delorme, R. (2020). Machine Learning for the Efficient Operational Optimization of Energy systems. Master's thesis, RWTH Aachen University.

Geers, P. (2020). Methoden zur Beschleunigung der Betriebsoptimierung von Energiesystemen (in German). Master's thesis, RWTH Aachen University.

Holtwerth, A. (2019). Automated data-driven model generation for energy systems using multivariate regression. Master's thesis, RWTH Aachen University.

The author additionally supervised the following student theses:

Theisen, M. (2020). Process monitoring for fault detection using sparse PCA algorithms. Project thesis, University of California, Davis, USA.

Peiler, L. (2019). Automatisierte Modellierung von Energiesystemen mittels neuronaler Netze (in German). Bachelor's thesis, RWTH Aachen University.

Reich, J. (2018). Entwicklung eines Verfahrens zur automatisierten Parametrierung von Einspeiseanlagen und Lasten auf Basis von maschinellem Lernen (in German). Master's thesis, RWTH Aachen University.

The work of all students is gratefully acknowledged.

Bibliography

Achterberg, T. and Wunderling, R. (2013). Mixed Integer Programming: Analyzing 12 Years of Progress. In Jünger, M. and Reinelt, G., editors, *Facets of Combinatorial Optimization: Festschrift for Martin Grötschel*, pages 449–481. Springer Berlin Heidelberg, Berlin, Heidelberg.

Akaike, H. (1974). A new look at the statistical model identification. *IEEE Transactions on Automatic Control*, 19(6):716–723.

Alabi, T. M., Aghimien, E. I., Agbajor, F. D., Yang, Z., Lu, L., Adeoye, A. R., and Gopaluni, B. (2022a). A review on the integrated optimization techniques and machine learning approaches for modeling, prediction, and decision making on integrated energy systems. *Renewable Energy*, 194:822–849.

Alabi, T. M., Lu, L., and Yang, Z. (2022b). Data-driven optimal scheduling of multi-energy system virtual power plant (MEVPP) incorporating carbon capture system (CCS), electric vehicle flexibility, and clean energy marketer (CEM) strategy. *Applied Energy*, 314:118997.

Alvarez, A. M., Louveaux, Q., and Wehenkel, L. (2017). A Machine Learning-Based Approximation of Strong Branching. *INFORMS Journal on Computing*, 29(1):185–195.

Anderson, R., Huchette, J., Ma, W., Tjandraatmadja, C., and Vielma, J. P. (2020). Strong mixed-integer programming formulations for trained neural networks. *Mathematical Programming*, 183(1):3–39.

Andiappan, V. (2017). State-Of-The-Art Review of Mathematical Optimisation Approaches for Synthesis of Energy Systems. *Process Integration and Optimization for Sustainability*, 1(3).

Awad, M. and Khanna, R. (2015). Support Vector Regression. In *Efficient Learning Machines*, pages 67–80. Apress, Berkeley, CA.

Bahl, B., Shu, D., Hollermann, D. E., Lützow, J., Hennen, M., Lampe, M., and Bardow, A. (2018a). Rigorous synthesis of energy systems by decomposition via time-series aggregation. *Computers & Chemical Engineering*, 112:70–81.

Bahl, B., Söhler, T., Hennen, M., and Bardow, A. (2018b). Typical Periods for Two-Stage Synthesis by Time-Series Aggregation with Bounded Error in Objective Function. *Frontiers in Energy Research*, 5:35.

Barber, C. B., Dobkin, D. P., and Huhdanpaa, H. (1996). The quickhull algorithm for convex hulls. *ACM Transactions on Mathematical Software*, 22(4):469–483.

Baumgärtner, N., Bahl, B., Hennen, M., and Bardow, A. (2019). RiSES3: Rigorous Synthesis of Energy Supply and Storage Systems via time-series relaxation and aggregation. *Computers & Chemical Engineering*, 127:127–139.

Baumgärtner, N., Shu, D., Bahl, B., Hennen, M., Hollermann, D. E., and Bardow, A. (2020). DeLoop: Decomposition-based Long-term operational optimization of energy systems with time-coupling constraints. *Energy*, 198:117272.

Benders, J. F. (1962). Partitioning procedures for solving mixed-variables programming problems. *Numerische Mathematik*, 4(1):238–252.

Bengio, Y., Lodi, A., and Prouvost, A. (2021). Machine learning for combinatorial optimization: A methodological tour d'horizon. *European Journal of Operational Research*, 290(2):405–421.

Bergner, M., Caprara, A., Ceselli, A., Furini, F., Lübbecke, M. E., Malaguti, E., and Traversi, E. (2015). Automatic Dantzig–Wolfe reformulation of mixed integer programs. *Mathematical Programming*, 149(1):391–424.

Bischi, A., Taccari, L., Martelli, E., Amaldi, E., Manzolini, G., Silva, P., Campanari, S., and Macchi, E. (2014). A detailed MILP optimization model for combined cooling, heat and power system operation planning. *Energy*, 74:12–26.

Bischi, A., Taccari, L., Martelli, E., Amaldi, E., Manzolini, G., Silva, P., Campanari, S., and Macchi, E. (2017). A rolling-horizon optimization algorithm for the long term operational scheduling of cogeneration systems. *Energy*, 184:73–90.

Bishop, C. M. (2016). *Pattern Recognition and Machine Learning*. Information Science and Statistics. Springer New York, New York, NY, softcover reprint of the original 1st edition 2006 (corrected at 8th printing 2009) edition.

Bixby, R. (2020). Mathematical Optimization: Past, Present, and Future (Part 2), `https://www.gurobi.com/resource/mathematical-optimization-past-present-and-future-part-2/`, accessed August 17, 2022.

Bonami, P., Lodi, A., and Zarpellon, G. (2018). Learning a Classification of Mixed-Integer Quadratic Programming Problems. In van Hoeve, W.-J., editor, *Integration of Constraint Programming, Artificial Intelligence, and Operations Research*, pages 595–604, Cham. Springer International Publishing.

Bonvin, D., Georgakis, C., Pantelides, C. C., Barolo, M., Grover, M. A., Rodrigues, D., Schneider, R., and Dochain, D. (2016). Linking models and experiments. *Industrial & Engineering Chemistry Research*, 55(25):6891–6903.

Box, G. (1979). Robustness in the Strategy of Scientific Model Building. In Launer, R. L. and Wilkinson, G. N., editors, *Robustness in Statistics*, pages 201–236. Academic Press.

Breiman, L. (1993). Hinging hyperplanes for regression, classification, and function approximation. *IEEE Transactions on Information Theory*, 39(3):999–1013.

Bruno, J. C., Fernandez, F., Castells, F., and Grossmann, I. E. (1998). A Rigorous MINLP Model for the Optimal Synthesis and Operation of Utility Plants. *Chemical Engineering Research and Design*, 76(3):246–258.

Burnham, K. P. and Anderson, D. R. (2003). *Model selection and multimodel inference: a practical information-theoretic approach.* Springer Science & Business Media.

Camponogara, E. and Nazari, L. F. (2015). Models and Algorithms for Optimal Piecewise-Linear Function Approximation. *Mathematical Problems in Engineering*, 2015:9.

Cao, K.-K., von Krbek, K., Wetzel, M., Cebulla, F., and Schreck, S. (2019). Classification and Evaluation of Concepts for Improving the Performance of Applied Energy System Optimization Models. *Energies*, 12(24):4656.

Cathcart, W. (2020). `https://twitter.com/wcathcart/status/1321949078381453314`, accessed on January 7, 2022.

Chilès, J.-P. and Delfiner, P. (2012). Kriging. In Shewhart, W. A. and Wilks, S. S., editors, *Geostatistics*, Wiley Series in Probability and Statistics, pages 147–237. John Wiley & Sons, Inc, Hoboken, NJ, USA.

Conejo, A. J., Castillo, E., García-Bertrand, R., and Mínguez, R. (2006). *Decomposition Techniques in Mathematical Programming: Engineering and Science Applications*. Springer-Verlag Berlin Heidelberg, Berlin, Heidelberg.

Cozad, A., Sahinidis, N. V., and Miller, D. C. (2014). Learning surrogate models for simulation-based optimization. *AIChE Journal*, 60(6):2211–2227.

Dantzig, G. B. and Wolfe, P. (1960). Decomposition Principle for Linear Programs. *Operations Research*, 8(1):101–111.

DeCarolis, J., Daly, H., Dodds, P., Keppo, I., Li, F., McDowall, W., Pye, S., Strachan, N., Trutnevyte, E., Usher, W., Winning, M., Yeh, S., and Zeyringer, M. (2017). Formalizing best practice for energy system optimization modelling. *Applied Energy*, 194:184–198.

Demirhan, C. D., Tso, W. W., Ogumerem, G. S., and Pistikopoulos, E. N. (2019). Energy systems engineering - a guided tour. *BMC Chemical Engineering*, 1(1):1–19.

D'Errico, J. (2009). SLM - shape language modeling.

DIN EN 16247 (2012). Energy audits.

Efstratiadis, A., Dialynas, Y. G., Kozanis, S., and Koutsoyiannis, D. (2014). A multivariate stochastic model for the generation of synthetic time series at multiple time scales reproducing long-term persistence. *Environmental Modelling & Software*, 62:139–152.

Elia, J. A. and Floudas, C. A. (2014). Energy supply chain optimization of hybrid feedstock processes: a review. *Annual review of chemical and biomolecular engineering*, 5:147–179.

Ernst, S. (1998). Hinging hyperplane trees for approximation and identification. In *Proceedings of the 37th IEEE 1998 Conference on Decision and Control (Cat. No.98CH36171)*, pages 1266–1271. IEEE.

Fernández, A., García, S., Galar, M., Prati, R. C., Krawczyk, B., and Herrera, F. (2018). *Learning from Imbalanced Data Sets*. Springer International Publishing, Cham.

Fischetti, M. and Jo, J. (2018). Deep neural networks and mixed integer linear optimization. *Constraints*, 23(3):296–309.

Floudas, C. A. (1995). *Nonlinear and mixed-integer optimization: Fundamentals and applications*. Topics in chemical engineering. Oxford University Press, New York and Oxford.

Frangopoulos, C. A. (2018). Recent developments and trends in optimization of energy systems. *Energy*, 164:1011–1020.

GAMS Development Corporation (2016). General Algebraic Modeling System (GAMS) Release 24.7.3.

Gao, X., Feng, Z., Wang, Y., Huang, X., Huang, D., Chen, T., and Lian, X. (2018). Piecewise Linear Approximation Based MILP Method for PVC Plant Planning Optimization. *Industrial & Engineering Chemistry Research*, 57(4):1233–1244.

Gasse, M., Chetelat, D., Ferroni, N., Charlin, L., and Lodi, A. (2019). Exact Combinatorial Optimization with Graph Convolutional Neural Networks. In Wallach, H., Larochelle, H., Beygelzimer, A., Alché-Buc, F. d., Fox, E., and Garnett, R., editors, *Advances in Neural Information Processing Systems*, volume 32. Curran Associates, Inc.

Geißler, B. (2011). *Towards Globally Optimal Solutions For MINLPs by Discretization Techniques with Applications in Gas Network Optimization*. PhD thesis, Naturwissenschaftliche Fakultät der Friedrich-Alexander-Universität Erlangen-Nürnberg.

Geißler, B., Martin, A., Morsi, A., and Schewe, L. (2012). Using Piecewise Linear Functions for Solving MINLPs. In Lee, J. and Leyffer, S., editors, *Mixed Integer Nonlinear Programming*, pages 287–314. Springer, New York.

Geurts, P., Ernst, D., and Wehenkel, L. (2006). Extremely randomized trees. *Machine Learning*, 63(1):3–42.

Gkioulekas, I. and Papageorgiou, L. G. (2018). Piecewise Regression through the Akaike Information Criterion using Mathematical Programming. *IFAC-PapersOnLine*, 51(15):730–735.

Gleixner, A., Bastubbe, M., Eifler, L., Gally, T., Gamrath, G., Gottwald, R. L., Hendel, G., Hojny, C., Koch, T., Lübbecke, M. E., Maher, S. J., Miltenberger, M., Müller, B., Pfetsch, M. E., Puchert, C., Rehfeldt, D., Schlösser, F., Schubert, C., Serrano, F., Shinano, Y., Viernickel, J. M., Walter, M., Wegscheider, F., Witt, J. T., and Witzig, J. (2018). The SCIP Optimization Suite 6.0: Technical Report.

Glorot, X., Bordes, A., and Bengio, Y. (2011). Deep Sparse Rectifier Neural Networks. In *AISTATS*.

Goderbauer, S., Bahl, B., Voll, P., Lübbecke, M. E., Bardow, A., and Koster, A. M. (2016). An adaptive discretization MINLP algorithm for optimal synthesis of decentralized energy supply systems. *Computers & Chemical Engineering*, 95:38–48.

Goderbauer, S., Comis, M., and Willamowski, F. J. (2019). The synthesis problem of decentralized energy systems is strongly NP-hard. *Computers & Chemical Engineering*, 124:343–349.

Goodfellow, I., Bengio, Y., and Courville, A. (2016). *Deep Learning*. MIT Press.

Graves, A. (2012). *Supervised Sequence Labelling with Recurrent Neural Networks*. Springer Berlin Heidelberg.

Grimstad, B. and Andersson, H. (2019). ReLU Networks as Surrogate Models in Mixed-Integer Linear Programs.

Grossmann, I. E. (2012). Advances in mathematical programming models for enterprise-wide optimization. *Computers & Chemical Engineering*, 47:2–18.

Guelpa, E., Bischi, A., Verda, V., Chertkov, M., and Lund, H. (2019). Towards future infrastructures for sustainable multi-energy systems: A review. *Energy*, 184:2–21.

Gupta, D., Maravelias, C. T., and Wassick, J. M. (2016). From rescheduling to online scheduling. *Chemical Engineering Research and Design*, 116:83–97.

Gurobi Optimization, L. L. (2020). Gurobi Optimizer Reference Manual.

Hochreiter, S. and Schmidhuber, J. (1997). Long short-term memory. *Neural computation*, 9(8):1735–1780.

Hornik, K., Stinchcombe, M., and White, H. (1989). Multilayer feedforward networks are universal approximators. *Neural Networks*, 2(5):359–366.

Huang, X., Xu, J., and Wang, S. (2010). Identification algorithm for standard continuous piecewise linear neural network. In *Proceedings of the 2010 American Control Conference*, pages 4931–4936.

Hurvich, C. M. and Tsai, C.-L. (1993). A corrected Akaike Information Criterion for vector autoregressive model selection. *Journal of Time Series Analysis*, 14(3):271–279.

IBM Corporation (2017). IBM ILOG CPLEX Optimization Studio.

IPCC (2013). Climate Change 2013: The Physical Science Basis: Contribution of Working Group I to the Fifth Assessment Report of the Intergovernmental Panel on Climate Change.

IPCC (2014). Climate change 2014: Mitigation of climate change: Working Group III contribution to the Fifth Assessment Report of the Intergovernmental Panel on Climate Change.

IPCC (2018). Global Warming of 1.5°C. An IPCC Special Report on the impacts of global warming of 1.5°C above pre-industrial levels and related global greenhouse gas emission pathways, in the context of strengthening the global response to the threat of climate change, sustainable development, and efforts to eradicate poverty.

ISO 50001:2018 (2018). Energy management systems – Requirements with guidance for use.

Kallrath, J. (2000). Mixed Integer Optimization in the Chemical Process Industry. *Chemical Engineering Research and Design*, 78(6):809–822.

Kämper, A., Delorme, R., Leenders, L., and Bardow, A. (2023). Boosting operational optimization of multi-energy systems by artificial neural nets. *Computers & Chemical Engineering*, 173:108208.

Kämper, A., Geers, P., Leenders, L., and Bardow, A. (2021a). Adaptive Rolling Horizon for operational optimization of multi-energy systems. In *Proceedings of the ECOS 2021 - 34th International Conference on Efficiency, Cost, Optimization, Simulation and Environmental Impact of Energy Systems (ECOS 2021).*

Kämper, A., Holtwerth, A., Leenders, L., and Bardow, A. (2021b). AutoMoG 3D: Automated Data-Driven Model Generation of Multi-Energy Systems Using Hinging Hyperplanes. *Frontiers in Energy Research*, 9:719658.

Kämper, A., Leenders, L., Bahl, B., and Bardow, A. (2021c). AutoMoG: Automated data-driven Model Generation of multi-energy systems using piecewise-linear regression. *Computers & Chemical Engineering*, 145:107162.

Kämper, A., Schricker, H., Böning, D., Hayen, N., Holtwerth, A., Bahl, B., Leenders, L., and Bardow, A. (2021d). AutoMoG 3D Automated data-driven modeling and optimization of multi-energy systems, GitLab Repository, `https://git-ce.rwth-aachen.de/ltt/automog-3d`.

Kantor, I., Robineau, J.-L., Bütün, H., and Maréchal, F. (2020). A Mixed-Integer Linear Programming Formulation for Optimizing Multi-Scale Material and Energy Integration. *Frontiers in Energy Research*, 8:49.

Katz, J., Pappas, I., Avraamidou, S., and Pistikopoulos, E. N. (2020). Integrating deep learning models and multiparametric programming. *Computers & Chemical Engineering*, 136:106801.

Kaufman, L. and Rousseeuw, P. (1987). *Clustering by means of medoids*. North-Holland: Reports of the Faculty of Mathematics and Informatics.

Kazda, K. and Li, X. (2021). Nonconvex Multivariate Piecewise-Linear Fitting Using the Difference-of-Convex Representation. *Computers & Chemical Engineering*, 106(7):107310.

Kenesei, T. and Abonyi, J. (2015). *Interpretability of Computational Intelligence-Based Regression Models*. SpringerBriefs in Computer Science. Springer International Publishing, Cham.

Kingma, D. P. and Ba, J. (2015). Adam: A Method for Stochastic Optimization. In Bengio, Y. and LeCun, Y., editors, *3rd International Conference on Learning Representations, ICLR 2015, San Diego, CA, USA, May 7-9, 2015, Conference Track Proceedings*.

Klöckl, B. and Papaefthymiou, G. (2010). Multivariate time series models for studies on stochastic generators in power systems. *Electric Power Systems Research*, 80(3):265–276.

Kocis, G. R. and Grossmann, I. E. (1989). Computational experience with dicopt solving MINLP problems in process systems engineering. *Computers & Chemical Engineering*, 13(3):307–315.

Kong, L. and Maravelias, C. T. (2020). On the Derivation of Continuous Piecewise Linear Approximating Functions. *INFORMS Journal on Computing*, 32(3):531–546.

Kong, X., Xiao, J., Liu, D., Wu, J., Wang, C., and Shen, Y. (2020). Robust stochastic optimal dispatching method of multi-energy virtual power plant considering multiple uncertainties. *Applied Energy*, 279:115707.

Kopanos, G. M. and Pistikopoulos, E. N. (2014). Reactive Scheduling by a Multiparametric Programming Rolling Horizon Framework: A Case of a Network of

Combined Heat and Power Units. *Industrial & Engineering Chemistry Research*, 53(11):4366–4386.

Kruber, M., Lübbecke, M. E., and Parmentier, A. (2017). Learning When to Use a Decomposition. In Salvagnin, D. and Lombardi, M., editors, *Integration of AI and OR techniques in constraint programming*, LNCS sublibrary. SL 1, Theoretical computer science and general issues, Cham. Springer.

LeCun, Y., Bengio, Y., and Hinton, G. (2015). Deep learning. *Nature*, 521(7553):436–444.

Li, Y., O'Neill, Z., Zhang, L., Chen, J., Im, P., and DeGraw, J. (2021). Grey-box modeling and application for building energy simulations - A critical review. *Renewable and Sustainable Energy Reviews*, 146:111174.

Lin, W. C. and Tsai, C. F. (2020). Missing value imputation: a review and analysis of the literature (2006–2017). *Artificial Intelligence Review*, 53:1487–1509.

Lodi, A. and Zarpellon, G. (2017). On learning and branching: a survey. *TOP*, 25(2):207–236.

Lougee-Heimer, R. (2003). The Common Optimization INterface for Operations Research: Promoting open-source software in the operations research community. *IBM Journal of Research and Development*, 47(1):57–66.

Mancarella, P. (2014). MES (multi-energy systems): An overview of concepts and evaluation models. *Energy*, 65:1–17.

Mancarella, P., Andersson, G., Peças-Lopes, J. A., and Bell K. R. W. (2016). Modelling of integrated multi-energy systems: Drivers, requirements, and opportunities. In *2016 Power Systems Computation Conference (PSCC)*, pages 1–22.

Marquant, J. F., Evins, R., and Carmeliet, J. (2015). Reducing Computation Time with a Rolling Horizon Approach Applied to a MILP Formulation of Multiple Urban Energy Hub System. *Procedia Computer Science*, 51:2137–2146.

Masti, D. and Bemporad, A. (2019). Learning binary warm starts for multiparametric mixed-integer quadratic programming. In *2019 18th European Control Conference (ECC)*, pages 1494–1499.

McBride, K. and Sundmacher, K. (2019). Overview of Surrogate Modeling in Chemical Process Engineering. *Chemie Ingenieur Technik*, 91(3):228–239.

McKay, M. D., Beckman, R. J., and Conover, W. J. (2000). A Comparison of Three Methods for Selecting Values of Input Variables in the Analysis of Output from a Computer Code. *Technometrics*, 42(1):55–61.

Misener, R., Gounaris, C. E., and Floudas, C. A. (2009). Global Optimization of Gas Lifting Operations: A Comparative Study of Piecewise Linear Formulations. *Industrial & Engineering Chemistry Research*, 48(13):6098–6104.

Mitsos, A., Asprion, N., Floudas, C. A., Bortz, M., Baldea, M., Bonvin, D., Caspari, A., and Schäfer, P. (2018). Challenges in process optimization for new feedstocks and energy sources. *Computers & Chemical Engineering*, 113:209–221.

Moretti, L., Martelli, E., and Manzolini, G. (2020). An efficient robust optimization model for the unit commitment and dispatch of multi-energy systems and microgrids. *Applied Energy*, 261:113859.

Murphy, K. P. (2012). *Machine learning: A probabilistic perspective*. Adaptive computation and machine learning series. MIT Press.

Nair, V., Bartunov, S., Gimeno, F., Glehn, I. v., Lichocki, P., Lobov, I., O'Donoghue, B., Sonnerat, N., Tjandraatmadja, C., Wang, P., Addanki, R., Hapuarachchi, T., Keck, T., Keeling, J., Kohli, P., Ktena, I., Li, Y., Vinyals, O., and Zwols, Y. (2021). Solving Mixed Integer Programs Using Neural Networks.

Nasrolahpour, E., Kazempour, S. J., Zareipour, H., and Rosehart, W. D. (2016). Strategic Sizing of Energy Storage Facilities in Electricity Markets. *IEEE Transactions on Sustainable Energy*, 7(4):1462–1472.

Obermeier, A., Vollmer, N., Windmeier, C., Esche, E., and Repke, J.-U. (2021). Generation of linear-based surrogate models from non-linear functional relationships for use in scheduling formulation. *Computers & Chemical Engineering*, 146:107203.

O'Malley, T., Bursztein, E., Long, J., Chollet, F., Jin, H., Invernizzi, L., et al. (2019). Keras Tuner, `https://github.com/keras-team/keras-tuner`.

Pillai, I., Fumera, G., and Roli, F. (2013). Threshold optimisation for multi-label classifiers. *Pattern Recognition*, 46(7):2055–2065.

Pucar, P. and Sjöberg, J. (1998). On the hinge-finding algorithm for hinging hyperplanes. *IEEE Transactions on Information Theory*, 44(3):1310–1319.

Qin, C., Zhao, J., Chen, L., Liu, Y., and Wang, W. (2021). An adaptive piecewise linearized weighted directed graph for the modeling and operational optimization of integrated energy systems. *Energy*, page 122616.

Rahmaniani, R., Crainic, T. G., Gendreau, M., and Rei, W. (2017). The Benders decomposition algorithm: A literature review. *European Journal of Operational Research*, 259(3):801–817.

Read, J., Pfahringer, B., Holmes, G., and Frank, E. (2009). Classifier Chains for Multi-label Classification. In Buntine, W., Grobelnik, M., Mladenić, D., and Shawe-Taylor, J., editors, *Machine Learning and Knowledge Discovery in Databases*, volume 5782 of *Lecture Notes in Computer Science*, pages 254–269. Springer Berlin Heidelberg.

Rebennack, S. and Krasko, V. (2020). Piecewise Linear Function Fitting via Mixed-Integer Linear Programming. *INFORMS Journal on Computing*, 32(2):507–530.

Reinert, C., Schellhas, L., Mannhardt, J., Shu, D., Kämper, A., Baumgärtner, N., Deutz, S., and Bardow, A. (2022). SecMOD: An Open-Source Modular Framework Combining Multi-Sector System Optimization and Life-Cycle Assessment. *Frontiers in Energy Research*, 10:884525.

Roll, J., Bemporad, A., and Ljung, L. (2004). Identification of piecewise affine systems via mixed-integer programming. *Automatica*, 40(1):37–50.

Rong, A., Lahdelma, R., and Luh, P. B. (2008). Lagrangian relaxation based algorithm for trigeneration planning with storages. *European Journal of Operational Research*, 188(1):240–257.

Ruan, G., Zhong, H., Zhang, G., He, Y., Wang, X., and Pu, T. (2020). Review of learning-assisted power system optimization. *CSEE Journal of Power and Energy Systems*, 7.

Schweidtmann, A. M. and Mitsos, A. (2019). Deterministic Global Optimization with Artificial Neural Networks Embedded. *Journal of Optimization Theory and Applications*, 180(3):925–948.

Shin, S. and Zavala, V. M. (2020). Diffusing-Horizon Model Predictive Control.

Shu, D., Baumgärtner, N., Dahmen, M., Bau, U., and Bardow, A. (2019). Optimal operation of energy systems with long-term constraints by time-series aggregation in receding horizon optimization. In Stanek, W., Gładysz, P., Werle, S., and Adamczyk, W., editors, *Proceedings of the 32nd International Conference on Efficiency, Cost, Optimization, Simulation and Environmental Impact of Energy Systems (ECOS 2019)*, pages 1971–1980.

Smola, A. J. and Schölkopf, B. (2004). A tutorial on support vector regression. *Statistics and Computing*, 14(3):199–222.

Smolin, Y. Y., Lau, K. K., and Soroush, M. (2019). First-principles modeling for optimal design, operation, and integration of energy conversion and storage systems. *AIChE Journal*, 65(7):e16482.

Srivastava, N., Hinton, G., Krizhevsky, A., Sutskever, I., and Salakhutdinov, R. (2014). Dropout: A Simple Way to Prevent Neural Networks from Overfitting. *Journal of Machine Learning Research*, 15:1929–1958.

Stoica, P. and Selén, Y. (2004). Model-order selection: a review of information criterion rules. *IEEE Signal Processing Magazine*, 21(4):36–47.

Taheri, S., Jooshaki, M., and Moeini-Aghtaie, M. (2021). Long-term planning of integrated local energy systems using deep learning algorithms. *International Journal of Electrical Power & Energy Systems*, 129:106855.

Tawarmalani, M. and Sahinidis, N. V. (2005). A polyhedral branch-and-cut approach to global optimization. *Mathematical Programming*, 103(2):225–249.

The Association of German Engineers (2008). VDI 4608 Part 2: Energy systems - Combined heat and power - Allocation and evaluation.

Thie, N., Franken, M., Schwaeppe, H., Bottcher, L., Muller, C., Moser, A., Schumann, K., Vigo, D., Monaci, M., Paronuzzi, P., Punzo, A., Pozzi, M., Gordini, A., Cakirer, K. B., Acan, B., Desideri, U., and Bischi, A. (2020). Requirements for Integrated Planning of Multi-Energy Systems. In *6th IEEE International Energy Conference (ENERGYCon)*, pages 696–701.

Vinyals, O., Fortunato, M., and Jaitly, N. (2015). Pointer Networks. In Cortes, C., Lawrence, N., Lee, D., Sugiyama, M., and Garnett, R., editors, *Advances in Neural Information Processing Systems*.

Voll, P. (2014). *Automated optimization-based synthesis of distributed energy supply systems: Aachen, Techn. Hochsch., Diss., 2013*, volume 1 of *Aachener Beiträge zur Technischen Thermodynamik*. Hochschulbibliothek der Rheinisch-Westfälischen Technischen Hochschule Aachen, Aachen.

Voll, P., Klaffke, C., Hennen, M., and Bardow, A. (2013). Automated superstructure-based synthesis and optimization of distributed energy supply systems. *Energy*, 50:374–388.

Wahab, A. S., Liew, P. Y., and Manaf, N. A. (2021). A Review on Data Reconciliation and Gross Error Detection for Process Plant Energy Management. *IOP Conference Series: Materials Science and Engineering*, 1051.

Wang, Q., McCalley, J. D., Zheng, T., and Litvinov, E. (2016). Solving corrective risk-based security-constrained optimal power flow with Lagrangian relaxation and Benders decomposition. *International Journal of Electrical Power & Energy Systems*, 75:255–264.

Wang, X., Palazoglu, A., and El-Farra, N. H. (2015). Operational optimization and demand response of hybrid renewable energy systems. *Applied Energy*, 143:324–335.

Weisberg, S. (2005). *Applied Linear Regression*. John Wiley & Sons.

Welsch, M., Mentis, D., and Howells, M. (2014). Long-Term Energy Systems Planning. In Jones, L. E., editor, *Renewable energy integration*, pages 215–225. Academic Press, London.

Wilson, Z. T. and Sahinidis, N. V. (2017). The ALAMO approach to machine learning. *Computers & Chemical Engineering*, 106:785–795.

Xavier, Á. S., Qiu, F., and Ahmed, S. (2020). Learning to Solve Large-Scale Security-Constrained Unit Commitment Problems. *INFORMS Journal on Computing*, 33(2):739-756.

Yang, D., Balaprakash, P., and Leyffer, S. (2021). Modeling Design and Control Problems Involving Neural Network Surrogates.

Yang, L., Liu, S., Tsoka, S., and Papageorgiou, L. G. (2016). Mathematical programming for piecewise linear regression analysis. *Expert Systems with Applications*, 44:156–167.

Yang, Y. and Wu, L. (2021). Machine learning approaches to the unit commitment problem: Current trends, emerging challenges, and new strategies. *The Electricity Journal*, 34(1):106889.

Yokoyama, R. and Ito, K. (1996). A Revised Decomposition Method for MILP Problems and Its Application to Operational Planning of Thermal Storage Systems. *Journal of Energy Resources Technology*, 118(4):277–284.

YouTube (2022). `https://blog.youtube/press/`, accessed on January 7, 2022.

Zarpellon, G., Jo, J., Lodi, A., and Bengio, Y. (2021). Parameterizing Branch-and-Bound Search Trees to Learn Branching Policies. In *Proceedings of the AAAI Conference on Artificial Intelligence*. AAAI Press.

Zhang, Q., Grossmann, I. E., Sundaramoorthy, A., and Pinto, J. M. (2016). Data-driven construction of Convex Region Surrogate models. *Optimization and Engineering*, 17(2):289–332.

Zhou, K., Kılınç, M. R., Chen, X., and Sahinidis, N. V. (2018). An efficient strategy for the activation of MIP relaxations in a multicore global MINLP solver. *Journal of Global Optimization*, 70(3):497–516.

Zhou, S., Di He, Zhang, Z., Wu, Z., Gu, W., Li, J., Li, Z., and Wu, G. (2019). A Data-Driven Scheduling Approach for Hydrogen Penetrated Energy System Using LSTM Network. *Sustainability*, 11(23):6784.

Aachener Beiträge zur Technischen Thermodynamik

ABTT 1
Philip Voll
Automated Optimization-Based Synthesis of Distributed Energy Supply Systems
1. Auflage 2014
ISBN 978-3-86130-474-6

ABTT 2
Johannes Jung
Comparative Life Cycle Assessment of Industrial Multi-Product Processes
1. Auflage 2014
ISBN 978-3-86130-471-5

ABTT 3
Franz Lanzerath
Modellgestützte Entwicklung von Adsorptionswärmepumpen
1. Auflage 2014
ISBN 978-3-86130-472-2

ABTT 4
Thorsten Brands
Einfluss der Gemischzusammensetzung auf die Verbrennung im Diesel- und GCAI-Motor
1. Auflage 2014
ISBN 978-3-95886-006-3

ABTT 5
Dominique Dechambre
Efficient Measurement of Liquid-Liquid Equilibria using Automation and Optimal Experimental Design
1. Auflage 2016
ISBN 978-395886-077-3

ABTT 6
Niklas von der Aßen
From Life-Cycle Assesement towards life-Cycle Design of Carbon Dioxide Capture and Utilization
1. Auflage 2016
ISBN 978-3-95886-080-3

ABTT 7
Matthias Lampe
Integrated Process and Organic Rankine Cycle Working Fluid Design in the Continuous-Molecular Targeting Framework
1. Auflage 2016
ISBN 978-3-95886-086-5

ABTT 8
Thomas Hülser
Optische Untersuchung der Zündvorgänge und deren Auswirkung auf die Verbrennung in PKW-Motoren
1. Auflage 2016
ISBN 978-3-95886-090-2

Aachener Beiträge zur Technischen Thermodynamik

ABTT 9
Malte Döntgen
Reaction Models from Reactive Molecular Dynamics and High-Level Kinetics Predictions
1. Auflage 2016
ISBN 978-3-95886-156-5

ABTT 10
Heike Schreiber
Experiments and Validated Models for Adsorption Thermal Energy Storage in Industrial and Residential Application
1. Auflage 2017
ISBN 978-3-95886-178-7

ABTT 11
André Dirk Sternberg
System-Wide Perspective for Life Cycle Assesment of CO_2-based C1-Chemicals
1. Auflage 2017
ISBN 978-3-95886-193-0

ABTT 12
Uwe Bau
From Dynamic Simulation to Optimal Design and Control of Adsorption Energy Systems
1. Auflage 2018
ISBN 978-3-95886-216-6

ABTT 13
Christian Jens
Modellbasiertes Design von Produkt, Lösungsmittel und Prozess für die Ameisensäure-synthese aus CO_2 und H_2
1. Auflage 2018
ISBN 978-3-95886-231-9

ABTT 14
Jan David Scheffczyk
Integrated Computer-Aided Design of Molecules and Processes using COSMO-RS
1. Auflage 2018
ISBN 978-3-95886-236-4

ABTT 15
Björn Bahl
Optimization-Based Synthesis of Large-Scale Energy Systems by Time-Series Aggregation
1. Auflage 2018
ISBN 978-3-95886-240-1

Aachener Beiträge zur Technischen Thermodynamik

ABTT 16
Bastian Liebergesell
A Milliliter-Scale Setup for the Efficient Characterization of Multicomponent Vapor-Liquid Equilibria Using Raman Spectroscopy
1. Auflage 2018
ISBN 978-3-95886-247-0

ABTT 17
Stefan Wilhelm Graf
A Design Approach for Adsorption Energy Systems Integrating Dynamic Modeling with Small-Scale Experiments
1. Auflage 2018
ISBN 978-3-95886-258-6

ABTT 18
Sebastian Kaminski
Quantum-Mechanics-Based Prediction of SAFT Parameters for Non-Associating and Associating Molecules Containing Carbon, Hydrogen, Oxygen and Nitrogen
1. Auflage 2019
ISBN 978-3-95886-270-8

ABTT 19
Maike Renate Hennen
Decision Support for the Synthesis of Energy Systems by Analysis of the Near-Optimal Solution Space
1. Auflage 2019
ISBN 978-3-95886-277-7

ABTT 20
Peyman Yamin
COSMO-RS-Based Methods for Improved Modelling of Complex Chemical Systems
1. Auflage 2019
ISBN 978-3-95886-288-3

ABTT 21
Meltem Erdogan
Assessement of Adsorbents for Drying by Experiments and Dynamic Simulations
1. Auflage 2019
ISBN 978-3-95886-303-3

ABTT 22
Christian Schulz
SRS/LIF-Messungen zur Charakterisierung rußarmer dieselähnlicher Flammen von alternativen Kraftstoffen und n-Heptan
1. Auflage 2019
ISBN 978-3-91886-310-1

Aachener Beiträge zur Technischen Thermodynamik

ABTT 23
Peter Beumers
Physically-Based Models for the Analysis of Raman Spectra
1. Auflage 2019
ISBN 978-3-95886-319-4

ABTT 24
Arne Kätelhön
Technology Choice Model for Consequential Life Cycle Assessment
1. Auflage 2019
ISBN 978-3-95886-324-8

ABTT 25
Christine Peters
Measurement of Multicomponent Diffusion in Liquids Using Raman Microspectroscopy and Microfluidics
1. Auflage 2020
ISBN 978-3-95886-337-8

ABTT 26
Dinah Elena Hollermann
Reliable and Robust Optimal Design of Sustainable Energy Systems
1. Auflage 2020
ISBN 978-3-95886-346-0

ABTT 27
Thomas Raffius
Laserspektroskopische Analyse von selbstzündenden motorischen Einspritzstrahlen alternativer Biokraftstoffe
1. Auflage 2020
ISBN 978-3-95886-358-3

ABTT 28
Johannes Schilling
Integrated Thermo-Economic Design of Processes and Molecules Using PC-SAFT
1. Auflage 2020
ISBN 978-3-95886-368-2

ABTT 29
Nils Julius Baumgärtner
Optimization of Low-Carbon Energy Systems from Industrial to National Scale
1. Auflage 2020
ISBN 978-3-95886-385-9

ABTT 30
Ludger Wolff
From Model-based Experimental Design and Analysis of Diffusion and Liquid-Liquid Equilibria to Process Applications
1. Auflage 2020
ISBN 978-3-95886-402-3

Aachener Beiträge zur Technischen Thermodynamik

ABTT 31
Andrej Gibelhaus
A Model-based Framework for Optimal Systems Integration of Adsorption Chillers
1. Auflage 2021
ISBN 978-3-95886-406-1

ABTT 32
Jan Seiler
Debottlenecking the Evaporator in Water-Based Adsorption Chillers
1. Auflage 2021
ISBN 978-3-95886-407-8

ABTT 33
Leif Kröger
Prediction of Reaction Rate Constants for the Synthesis of Microgels
1. Auflage 2021
ISBN 978-395886-425-2

ABTT 34
Sarah von Pfingsten
Uncertainty Analysis in Matrix-Based Life Cycle Assessment
1. Auflage 2022
ISBN 978-3-95886-431-3

ABTT 35
Ludger Leenders
Optimization Methods for Integrating Energy and Production Systems
1. Auflage 20022
ISBN 978-3-95886-445-0

ABTT 36
Leonard Müller
Harmonized Life Cycle Assessment of Technologies for Carbon Capture and Utilization
1. Auflage 2022
ISBN 978-3-95886-434-4

ABTT 37
Fritz Röben
Decarbonization of Copper Production by Optimal Demand Response
and Power-to-Hydrogen
1. Auflage 2022
ISBN 978-3-95886-458-0

ABTT 38
Johanna Kleinekorte
Predictive Life Cycle Assessment for Chemical Processes using Machine Learing
1. Auflage 2022
ISBN 978-3-95886-461-0

Aachener Beiträge zur Technischen Thermodynamik

ABTT 39
Raoul Meys
Designing Pathways for Net-Zero Greenhouse Gas Emission Plastics with Life Cycle Optimization
1. Auflage 2022
ISBN 978-3-95886-463-4

ABTT 40
Lukas Krep
Novel Acceleration Methods and Improved Transition State Finding Approaches for the Automatic Exploration of Reaction Networks
1. Auflage 2023
ISBN 978-3-95886-480-1

ABTT 41
Andreas Kämper
Data-driven Modeling and Optimization of Multi-Energy Systems
1. Auflage 2023
ISBN 978-3-95886-488-7